P9-BUI-954

Everyday Mathematics®

The University of Chicago School Mathematics Project

Student Reference Book

Everyday Mathematics®

The University of Chicago School Mathematics Project

Student Reference Book

The McGraw·Hill Companies

UCSMP Elementary Materials Component
Max Bell, Director

Authors
Max Bell, Jean Bell, John Bretzlauf, Amy Dillard, Robert Hartfield, Andy Isaacs, James McBride, Kathleen Pitvorec, Peter Saecker

Technical Art
Diana Barrie

Assistants
Lance Campbell (Research), Adam Fischer (Editorial), James Flanders (Technology), Deborah Arron Leslie (Research), John Saller (Research)

www.WrightGroup.com

Printed in the United States of America.

Send all inquiries to:
Wright Group/McGraw-Hill
P.O. Box 812960
Chicago, IL 60681

ISBN 0-07-604569-2

5 6 7 8 9 QWD 11 10 09 08 07

The McGraw·Hill Companies

Contents

Operations and Computation 49

Data and Chance 75

Geometry 95

Measurement 131

Data Bank 211

Problem Solving 249

Calculators 261

Games 267

About the *Student Reference* Book

A reference book is a book that is organized to help people find information quickly and easily. Some reference books that you may have used are dictionaries, encyclopedias, cookbooks, and even phone books.

You can use this *Student Reference Book* to look up and review topics in mathematics. It has the following sections:

- A **Table of Contents** that lists the sections and shows how the book is organized. Each section has the same color band across the top of the pages.
- **Essays** within each section, such as Number Lines, Negative Numbers, Trade-First Subtraction Method, Bar Graphs, Triangles, Perimeter, Calendars, Adjusting Numbers, Number Patterns, and Solving Number Stories.
- A collection of **photo essays** called **Mathematics... Every Day,** that show in words and pictures some of the ways that mathematics is used.
- Directions on how to play **mathematical games** that help you practice math skills.
- A **Data Bank** with posters, maps, and other information.
- A **Glossary** of mathematical terms consisting of brief definitions of important words.
- An **Answer Key** for every Check Your Understanding problem in the book.
- An **Index** to help you locate information quickly.

Numbers and Counting

Number Uses

Most people use hundreds or thousands of numbers each day. There are numbers on clocks, calendars, car license plates, postage stamps, scales, and so on. These numbers are used in many different ways.

The major ways for using numbers are listed below.

1. Numbers are used as **counts.** A count is a number that tells "how many" things there are.

12 eggs 275 pages 24 keys

The **counting numbers** are the numbers used in counting: 1, 2, 3, 4, 5, and so on.

2. Numbers are used as **measures.** A measure is a number that tells "how much" of something there is. For example, a scale is used to measure how much something weighs. And a ruler is used to measure how much space there is between two points.

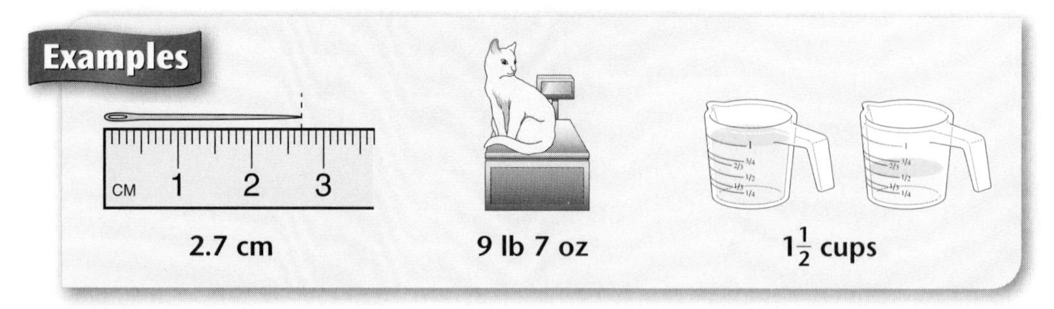

2.7 cm 9 lb 7 oz $1\frac{1}{2}$ cups

The numbers 1, 2, 3, and so on are the only numbers we need for counting. But for careful measuring we also need "in-between" numbers. The numbers between counting numbers are called **fractions** and **decimal numbers.**

For example, a ruler that shows only inch marks may not be useful for making accurate measurements. We need marks between the inch marks to show fractions of inches.

3. Numbers are used to show **locations** compared to some starting point.

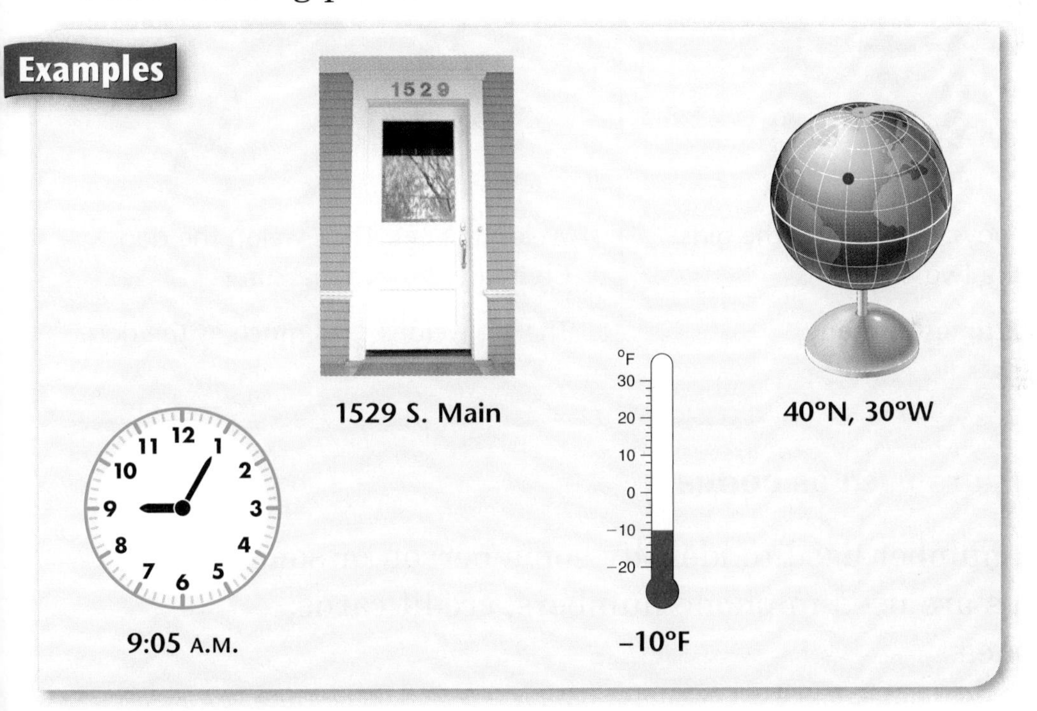

Numbers in the street address 1529 S. Main give a location on Main Street. Clock times are locations in time, starting at noon or midnight. A pair of numbers like 40°N, 30°W gives a location on Earth's surface compared to the location of the equator and the prime meridian.

Temperatures give a location on a thermometer starting at 0 degrees. We use negative numbers to show temperatures that are below 0 degrees. A temperature of $-10°F$ is read as "10 degrees below zero." The numbers -1, -2, -3, $-\frac{1}{2}$, and -31.6 are all negative numbers.

4. Numbers are used to make **comparisons.**

We often use numbers to compare two counts or measures.

Examples

Count the boys. Then count the girls. Compare the two counts.

There are **3 times as many** boys as girls.

Weigh the cat. Then weigh the dog. Compare the two measures.

The cat weighs $\frac{1}{3}$ **as much** as the dog.

5. Numbers are used as **codes.**

A code is a number used to identify some person or some thing. Codes are used in phone numbers, credit cards, and ZIP codes.

For example, in the ZIP code **60637:**

6 refers to the midwestern part of the United States.

06 refers to Chicago.

37 refers to a certain neighborhood in Chicago.

Examples

772-555-1212

phone number

M268-425-206

driver's license

bar code and ISBN code

Reminder: Most numbers come with a unit or symbol that tells what the number means: 10 cats, 10 inches, 10 A.M., and 10°F mean different things. The unit or symbol shows which meaning you want.

Did You Know?

A bar code identifies a product and the company that made it. When a bar code is scanned at a store, the bar code number is sent to the store's computer. The computer does a "price lookup" for that bar code number. And the price is shown on the cash register.

The ISBN code is used to identify books.

Check Your Understanding

Decide whether each number is used to count, to measure, to show a location, to compare two quantities, or as a code.

1. 25 meters

2. 1-800-555-1212

3. 13 ducks

4. 11/18/98

5. 4 times as much

6. 2 lb 5 oz

7. 237 Church St.

8. $2\frac{1}{4}$ inches

9. 303 children

10. $\frac{1}{2}$ as many

11. 12 noon

12. 25 minutes

13. 7:40 P.M.

14. 0 pink elephants

15. 100°C

16. 9:15 A.M.

Check your answers on page 335.

Number Grids

A monthly calendar is an example of a **number grid.** The numbers of the days of the month are listed in order.

The numbers are printed in boxes. The boxes are printed in rows. There are 7 boxes in each row because there are 7 days in a week.

May						
Sun	Mon	Tue	Wed	Thu	Fri	Sat
				1	2	3
4	5	6	7	8	9	10
11	12	13	14	15	16	17
18	19	20	21	22	23	24
25	26	27	28	29	30	31

A number grid like the one you use in class is shown on the next page. The numbers are listed in order and printed in rows of boxes.

Counting forward on a number grid is like reading a calendar. When you reach the end of a line, you go to the next line below and start at the left. Counting backward on a number grid is like reading a calendar backward.

Your class number grid has 10 boxes in each row. The number in the last box of each row has 0 as the final digit. This is the kind of number grid that we shall study.

–19	–18	–17	–16	–15	–14	–13	–12	–11	–10
–9	–8	–7	–6	–5	–4	–3	–2	–1	0
1	2	3	4	5	6	7	8	9	10
11	12	13	14	15	16	17	18	19	20
21	22	23	24	25	26	27	28	29	30
31	32	33	34	35	36	37	38	39	40
41	42	43	44	45	46	47	48	49	50
51	52	53	54	55	56	57	58	59	60
61	62	63	64	65	66	67	68	69	70
71	72	73	74	75	76	77	78	79	80
81	82	83	84	85	86	87	88	89	90
91	92	93	94	95	96	97	98	99	100
101	102	103	104	105	106	107	108	109	110
111	112	113	114	115	116	117	118	119	120

The numbers on a number grid have some simple patterns. These patterns make the grid easy to use.

- When you move *right,* numbers *increase by 1.* (16 is 1 more than 15.)
- When you move *left,* numbers *decrease by 1.* (23 is 1 less than 24.)
- When you move *down,* numbers *increase by 10.* (75 is 10 more than 65.)
- When you move *up,* numbers *decrease by 10.* (91 is 10 less than 101.)

Example Part of a number grid is shown below.

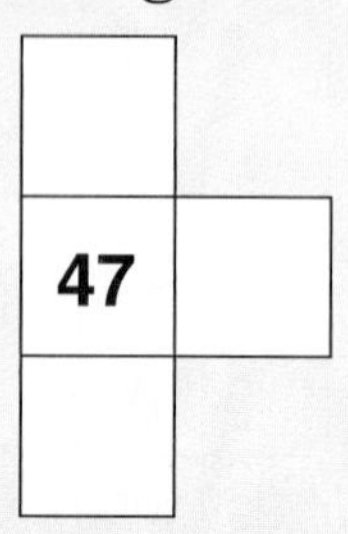

You can use number-grid patterns to fill in the missing numbers.

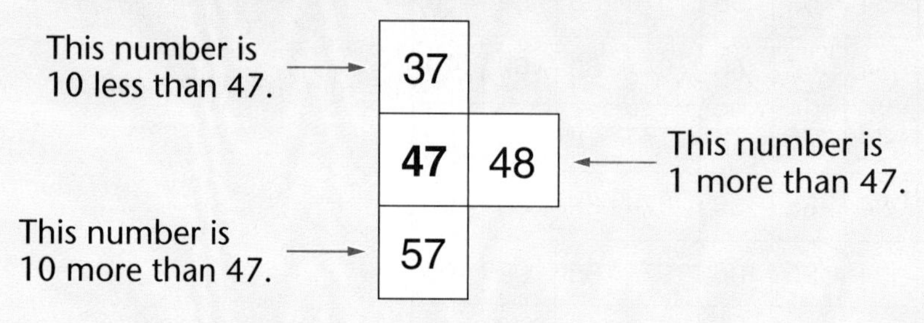

A number grid can help you find the difference between two numbers.

Example Find the difference between 37 and 64.

31	32	33	34	35	36	37	38	39	40
41	42	43	44	45	46	47	48	49	50
51	52	53	54	55	56	57	58	59	60
61	62	63	64	65	66	67	68	69	70

- Start at 37.
- Count the number of *tens* going down to 57. There are 2 tens, or 20.
- Count the number of *ones* going right from 57 to 64. There are 7 ones, or 7.
- The difference between 37 and 64 is 2 tens and 7 ones, or 27.

Number grids can also be used to explore number patterns.

Example Start with 0. Count by 2s until you reach 100.

The blue boxes contain *even* numbers.

The orange boxes contain *odd* numbers.

									0
1	2	3	4	5	6	7	8	9	10
11	12	13	14	15	16	17	18	19	20
21	22	23	24	25	26	27	28	29	30
31	32	33	34	35	36	37	38	39	40
41	42	43	44	45	46	47	48	49	50
51	52	53	54	55	56	57	58	59	60
61	62	63	64	65	66	67	68	69	70
71	72	73	74	75	76	77	78	79	80
81	82	83	84	85	86	87	88	89	90
91	92	93	94	95	96	97	98	99	100

Check Your Understanding

1. Use the number grid above to find the difference.

   a. Between 16 and 46 **b.** Between 73 and 98

   c. Between 37 and 72

2. Copy the parts of the number grids shown. Use number-grid patterns to find the missing numbers.

a.

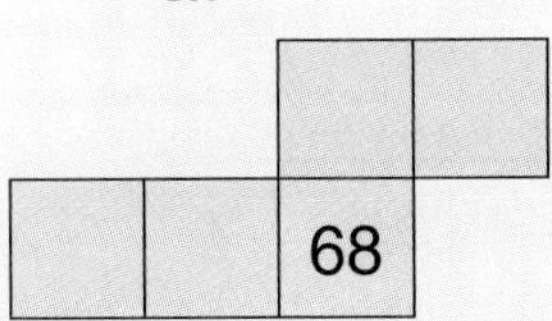

b.

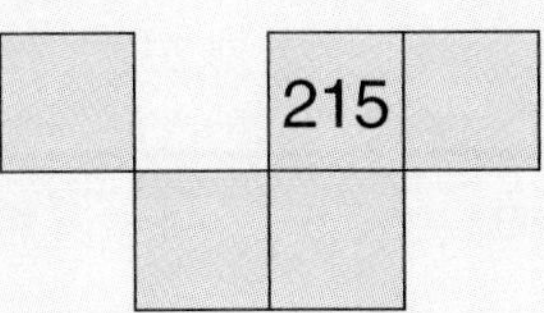

c.

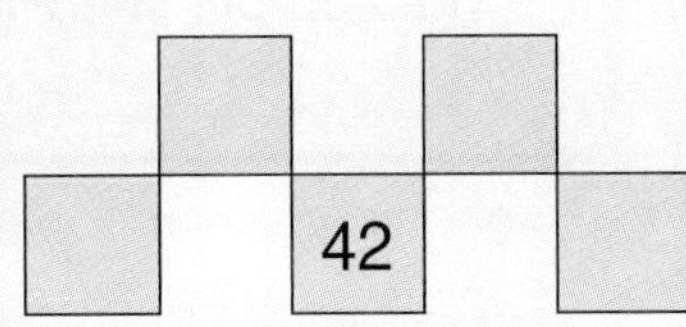

Check your answers on page 335.

Number Lines

A **number line** is a line with numbers marked beside it. An example is shown below.

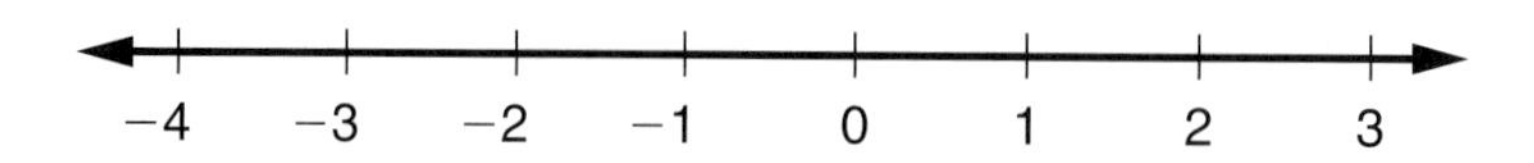

The number 0 is called the **zero point.** All of the spaces between marks are the same length.

The numbers to the right of 0 are called **positive numbers.** The numbers to the left of 0 are called **negative numbers.** For example, –3 is called "negative 3."

Example The number line below shows the numbers –2, –1, 0, 1, 2, and 3.

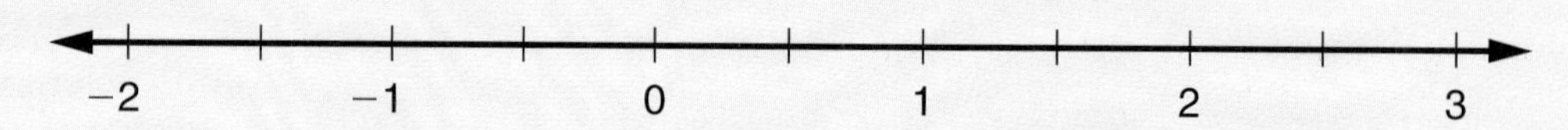

Marks have been drawn halfway between the numbered marks.

We can use fractions to write in the numbers for these halfway marks. The number line will look like this:

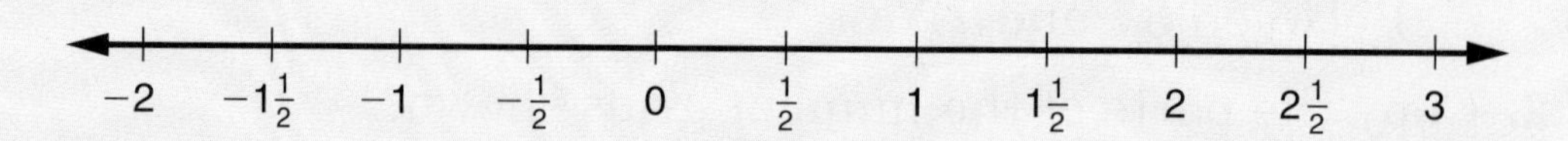

Or, we can use decimals to write in numbers for the halfway marks. The number line will look like this:

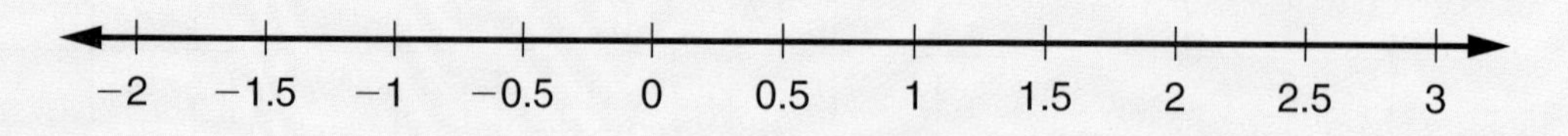

Example Every ruler is a number line. If the zero mark is at the end of the ruler, the number 0 may not be printed on the ruler.

On rulers, inches are usually divided into halves, quarters, eighths, and sixteenths. The marks to show fractions of an inch are usually of different sizes.

Example Every thermometer is a number line.

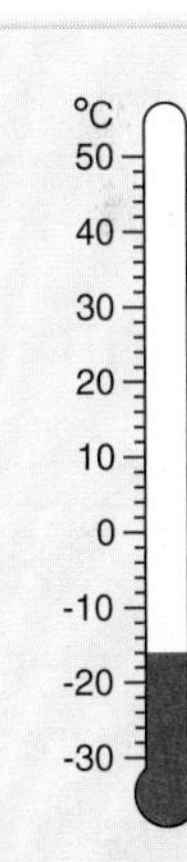

The zero mark on a Celsius scale (0°C) is the temperature at which water freezes.

Negative numbers are shown on the thermometer. A temperature of −16°C is read as "16 degrees below zero."

The marks on a thermometer are evenly spaced. The space between marks is usually 2 degrees.

Here is an easy method you can use to fill in the missing numbers on a number line:

1. Find the distance between the endpoints.
2. Count the number of spaces between the endpoints.
3. This fraction gives the length of each space:

$$\frac{\text{distance between endpoints}}{\text{number of spaces between endpoints}}$$

distance between endpoints ← numerator
number of spaces between endpoints ← denominator

Did You Know?

Forty degrees below zero is the only temperature at which Celsius and Fahrenheit thermometers will show the same reading. −40°C = −40°F

Example Fill in the number line. The endpoints are 72 and 73.

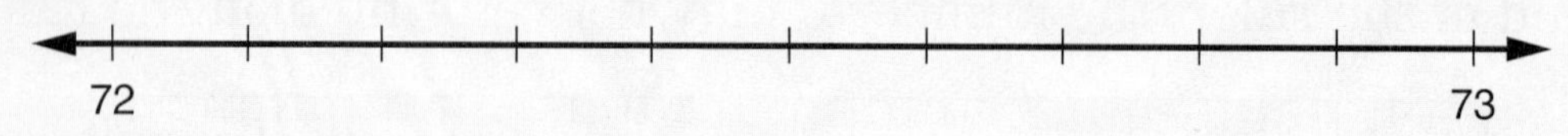

1. The distance from 72 to 73 is 1.
2. There are 10 spaces between 72 and 73.
3. Write the fraction $\frac{1}{10}$. The length of each space is $\frac{1}{10}$ or 0.1.

Here is the filled-in number line:

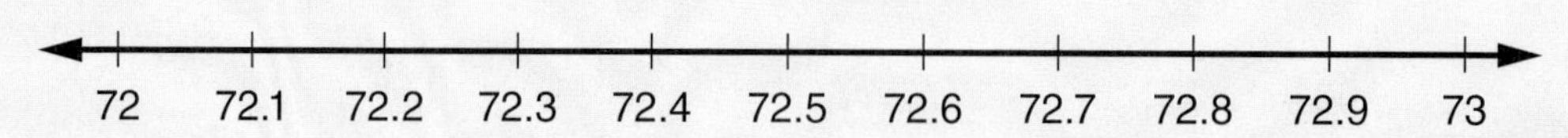

Example Fill in the number line. The endpoints are 200 and 204.

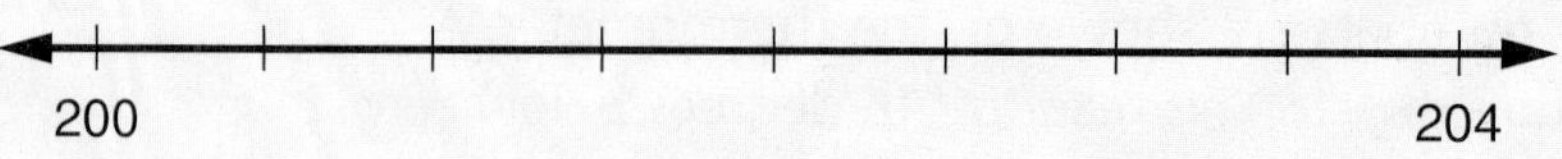

1. The distance from 200 to 204 is 4.
2. There are 8 spaces between 200 and 204.
3. Write the fraction $\frac{4}{8}$. $\frac{4}{8} = \frac{1}{2}$. The length of each space is $\frac{1}{2}$.

Here is the filled-in number line:

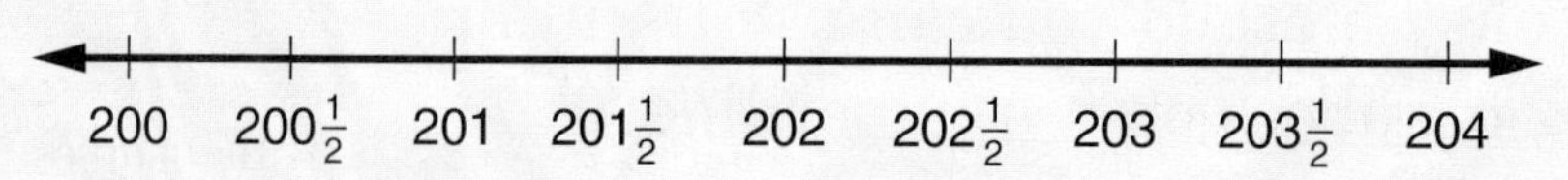

Check Your Understanding

Copy the number lines. Write the missing numbers.

1.

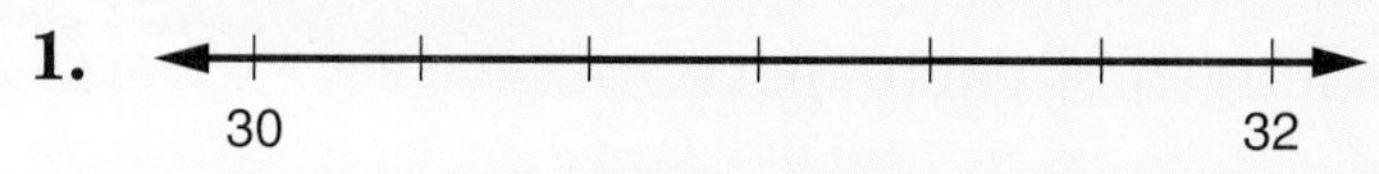

2.

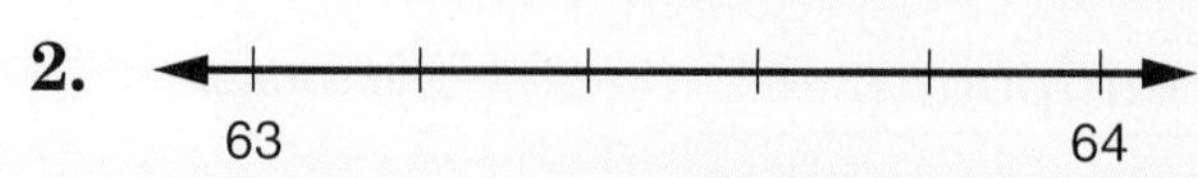

Check your answers on page 335.

Comparing Numbers

When two numbers are **compared,** two results are possible:

- The numbers are **equal.**
- The numbers are **not equal.**
  One of the numbers is larger than the other.

Different symbols are used to show that numbers are equal or not equal.

- Use an **equal sign** (=) to show that the numbers are *equal.*
- Use a **greater-than symbol** (>) or a **less-than symbol** (<) to show that the numbers are *not equal.*

Here is one way to remember the meaning of the > and < symbols. Think of each symbol as a mouth. The mouth must be open to swallow the larger number.

Examples The table below lists other examples. Some examples compare numbers, and others compare amounts.

Symbol	Meaning	Examples
$=$	"equals" "is the same as"	$20 = 4 \times 5$ 3 cm = 30 mm $\frac{1}{2} = 0.5$
$>$	"is greater than"	14 ft 7 in. > 13 ft 11 in. $1.23 > 1.2$
$<$	"is less than"	$2\frac{1}{2} < 4$ 8 thousand < 12,000,000

Name-Collection Boxes

Any number can be written in many different ways. Different names for the same number are called **equivalent names.**

A **name-collection box** is a place to write names for the same number. It is a box with an open top and a label attached to it.

- The name on the label gives a number.
- The names written inside the box are equivalent names for the name on the label.

Example A name-collection box for 8 is shown below. It is called an "8-box."

8

2×4 卌 |||

0.8×10 eight $8 - 0$

$8 \div 1$ *ocho*   • • / • • / • • / • •

$2 + 2 + 2 + 2$

To form equivalent names for numbers, you can

- add, subtract, multiply, or divide
- use tally marks or arrays
- write words in English or other languages

Check Your Understanding

Write five equivalent names for the number 12.

Check your answers on page 335.

Example A name-collection box for 50 is shown below. It is called a "50-box."

50		
	$100 \div 2$	5×10
$10 + 10 + 10 + 10 + 10$		
1 more than 49		$25 + 25$
fifty	*cincuenta*	

Each name in the 50-box is a different way to say the number 50. This means that we can use an equal sign (=) to write each statement below.

$50 = 5 \times 10$ $\quad 50 = 25 + 25$ $\quad 100 \div 2 = 50$ $\quad$ fifty $= 50$

Check Your Understanding

1. What name belongs on the label for this name-collection box?

?		
	$12 - 3$	$19 - 10$
𝍸 ||||	$\frac{90}{10}$	
	$4 + 5$	

2. Draw a 6-box like the one shown. Write five equivalent names for 6 in your 6-box.

Check your answers on page 335.

Parentheses

What does $15 - 3 + 2$ equal? Should you *add* or *subtract* first?

We use **parentheses ( )** in number problems to tell which operation to do first.

Example What does $(15 - 3) + 2$ equal?

The parentheses tell you to subtract first. $15 - 3 = 12$
Then add $12 + 2$.

The answer is 14.

Example What does $15 - (3 + 2)$ equal?

The parentheses tell you to add first. $3 + 2 = 5$
Then subtract $15 - 5$.

The answer is 10.

Example $5 \times (9 - 2) = ?$

The parentheses tell you to subtract first. $9 - 2 = 7$
Then multiply 5×7. This equals 35.

The answer is 35.

Example What does $(2 \times 3) + (4 \times 5)$ equal?

There are 2 sets of parentheses. Solve each problem that is inside parentheses first.
$2 \times 3 = 6$ and $4 \times 5 = 20$.
Then add these answers. $6 + 20 = 26$

The answer is 26.

Sometimes a number statement does not have parentheses. You are asked to add parentheses.

Example Make this number statement true by adding parentheses: $18 = 6 + 3 \times 4$.

There are two ways to add parentheses:

$18 = (6 + 3) \times 4$ and $18 = 6 + (3 \times 4)$

Add first. $6 + 3 = 9$
Then multiply. $9 \times 4 = 36$
18 is not equal to 36.
This statement is false.

Multiply first. $3 \times 4 = 12$
Then add. $6 + 12 = 18$
18 is equal to 18.
This statement is true.

The correct way to add parentheses is $18 = 6 + (3 \times 4)$.

Example Add parentheses to make this statement true: $14 - 6 \div 2 = 4$.

There are two ways to add parentheses:

$(14 - 6) \div 2 = 4$ and $14 - (6 \div 2) = 4$

Subtract first. $14 - 6 = 8$
Then divide. $8 \div 2 = 4$
4 is equal to 4.
This statement is true.

Divide first. $6 \div 2 = 3$
Then subtract. $14 - 3 = 11$
11 is not equal to 4.
This statement is false.

The correct way to add parentheses is $(14 - 6) \div 2 = 4$.

Check Your Understanding

Add parentheses to make each statement true.

1. $20 - 12 + 5 = 13$

2. $30 = 5 + 5 \times 5$

3. $4 \times 7 + 14 = 84$

4. $16 = 2 \times 3 + 1 \times 2$

Check your answers on page 335.

Place Value for Counting Numbers

People all over the world write numbers in the same way. This system of writing numbers was invented in India about 1,500 years ago. It is called a **place-value system.**

Any number can be written using the **digits** 0, 1, 2, 3, 4, 5, 6, 7, 8, and 9. The **place** that each digit has in a number is very important.

Example The numbers 72 and 27 use the same digits, a 7 and a 2. But 72 and 27 are different numbers because the 7 and the 2 are in different places.

tens place	ones place
7	2

The digit 2 in 72 is worth 2 (2 ones).
The digit 7 in 72 is worth 70 (7 tens).

tens place	ones place
2	7

The digit 2 in 27 is worth 20 (2 tens).
The digit 7 in 27 is worth 7 (7 ones).

Example The number 55 uses the digit 5 twice. But the two 5s are in different places.

The 5 in the tens place is worth 50 (5 tens).
The 5 in the ones place is worth 5 (5 ones).

tens place	ones place
5	5

Example The digit 0 is very important, even though it is not worth anything.

tens place	ones place
7	0

The digit 7 is worth 70 in this number.

tens place	ones place
	7

With the 0 removed, the digit 7 is worth only 7.

We can use a **place-value chart** to show how much each digit in a number is worth. The **place** for a digit is its position in the number. The **value** of a digit is how much it is worth.

Example The number 30,628 is shown in the place-value chart below.

10,000s	1,000s	100s	10s	1s
10 thousands place	thousands place	hundreds place	tens place	ones place
3	0	6	2	8

The digit 8 in the 1s place is worth $8 \times 1 = 8$.
The digit 2 in the 10s place is worth $2 \times 10 = 20$.
The digit 6 in the 100s place is worth $6 \times 100 = 600$.
The digit 0 in the 1,000s place is worth $0 \times 1,000 = 0$.
The digit 3 in the 10,000s place is worth $3 \times 10,000 = 30,000$.

30,628 is read as "thirty thousand, six hundred twenty-eight."

For larger numbers we can use larger place-value charts. Look for the commas that separate groups of 3 digits. The commas will help you identify the thousands, millions, and so on.

Example Read the number shown below.

Millions				**Thousands**				**Ones**		
hundred millions	ten millions	millions	,	hundred thousands	ten thousands	thousands	,	hundreds	tens	ones
		7	,	3	3	4	,	6	0	9

Read from left to right.
Read "million" at the first comma. Read "thousand" at the next comma.

The number is read as "7 **million,** 334 **thousand,** 609."

A place-value chart can be used to compare two numbers.

Example Compare the numbers 63,429 and 63,942. Which number is greater?

10,000s 10 thousands	**1,000s** thousands	**100s** hundreds	**10s** tens	**1s** ones
6	3	4	2	9
6	3	9	4	2

Start at the left side.
The 10,000s digits *are* the same. They are both worth 60,000.
The 1,000s digits *are* the same. They are both worth 3,000.
The 100s digits are *not* the same. The 4 is worth 400 and the 9 is worth 900.
The 9 is worth more.

So 63,942 is the larger number.

Example Phil and Meg use place-value charts to write the number 35.

Phil

10s tens	**1s** ones
3	5

Meg

10s tens	**1s** ones
2	15

Phil's chart shows 3 tens and 5 ones. This is 30 + 5 = 35.

Meg's chart shows 2 tens and 15 ones. This is 20 + 15 = 35.
Meg's way of filling in the place-value chart may be unusual, but it is correct.

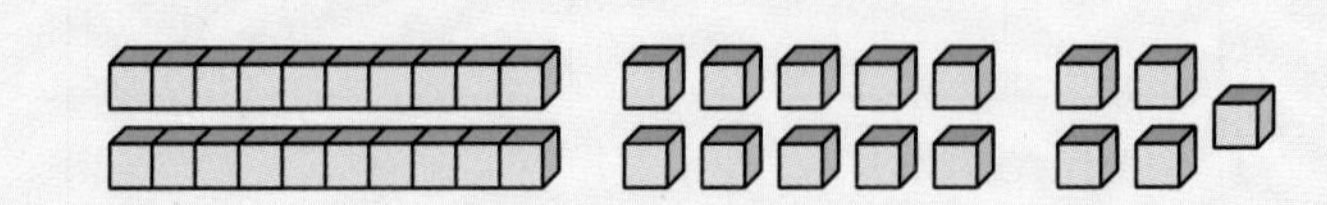

Phil and Meg are both correct.

Check Your Understanding

1. Count by 10s from 142. Write the next ten numbers.

2. Write the number that has

7 in the tens place,

8 in the 10 thousands place,

6 in the ones place,

0 in the hundreds place, and

2 in the thousands place.

3. Write the number that is 1,000 more.

a. 3,789

b. 7,890

4. Write the number that is 1,000 less.

a. 7,671

b. 2,874

5. What is the largest 3-digit number you can make using the digits 3, 7, and 5?

Check your answers on page 336.

Did You Know?

The Babylonians were the first people to use a symbol for a missing amount. They used the symbol shown here to stand for zero as early as 300 B.C.

Using Fractions to Name Part of a Whole

The numbers $\frac{1}{2}$, $\frac{3}{4}$, $\frac{5}{4}$, and $\frac{25}{100}$ are all **fractions.** A fraction is written with two numbers. The top number of a fraction is called the **numerator.** The bottom number is called the **denominator.**

When naming a fraction, we name the numerator first. Then we name the denominator.

three-fourths $\frac{3}{4}$ ← numerator → $\frac{25}{100}$ twenty-five hundreths
← denominator →

Fractions can be used to name part of a whole object.

Example What fraction of this square is shaded?

The whole object is the square.
It has been divided into 8 equal parts.
Each part is $\frac{1}{8}$ (one-eighth) of the square.
Three of the parts are shaded.

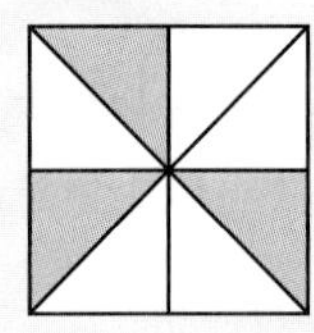

So $\frac{3}{8}$ (three-eighths) of the square is shaded.

3 ← The *numerator* 3 tells the number of *shaded* parts.
8 ← The *denominator* 8 tells the number of equal parts in the *whole* square.

Example What fraction of this circle is shaded?

The circle is divided into 12 equal parts.
Each part is $\frac{1}{12}$ (one-twelfth) of the circle.
Five of the parts are shaded.

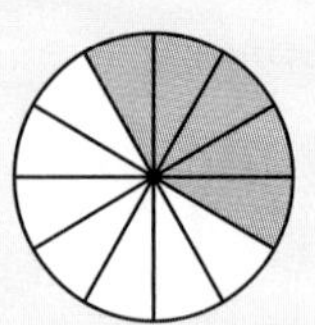

The shaded part is $\frac{5}{12}$ (five-twelfths) of the circle.

Examples What fraction names each shaded part of the rectangle?

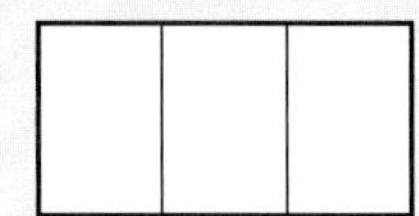

The shaded part is $\frac{0}{3}$ (zero-thirds) of the rectangle.

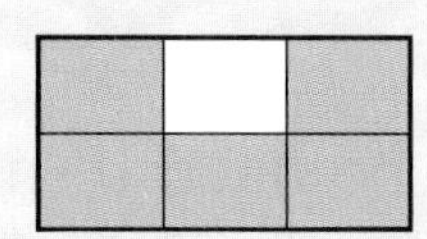

The shaded part is $\frac{5}{6}$ (five-sixths) of the rectangle.

When the numerator is zero, the fraction is equal to zero.

Examples $\frac{0}{2}, \frac{0}{8}, \frac{0}{12}, \frac{0}{100}$ Each of these fractions equals 0.

When the numerator and denominator are the same, the fraction is equal to 1.

Examples $\frac{2}{2}, \frac{8}{8}, \frac{12}{12}, \frac{100}{100}$ Each of these fractions equals 1.

Check Your Understanding

1. Write each of these fractions.

a. five-sixteenths **b.** two-thirds **c.** four-thirds

2. What fraction names each shaded part?
Write the fraction as a number and with words.

a. 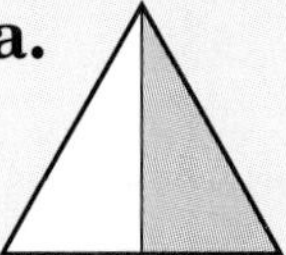

b. 

c. 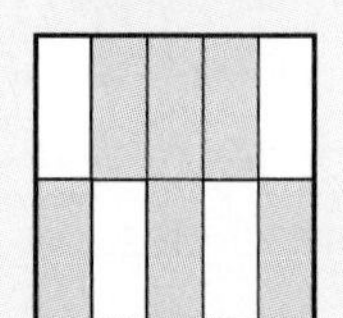

d. 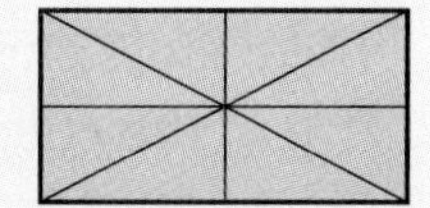

Check your answers on page 336.

Using Fractions to Name Part of a Collection

Fractions can be used to name part of a collection.

Example What fraction of the buttons are small?

There are 7 buttons in all.
Three buttons are small.
Three out of 7 buttons are small.
This fraction shows what part of the button collection is small buttons:

$\frac{3}{7}$ ← number of small buttons / ← number of buttons in all

$\frac{3}{7}$ (three-sevenths) of the buttons are small.

Example What fraction of the dots are circled?

There are 12 dots.
Seven dots are circled.
Seven out of 12 dots are circled.

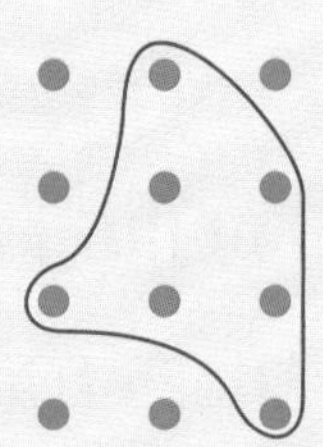

The fraction of dots circled is $\frac{7}{12}$.

Example What fraction of the counters are circles?

There are 12 counters in all.
Five of the counters are circles.
Five out of 12 counters are circles.

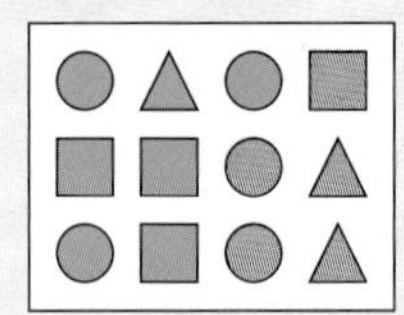

The fraction of counters that are circles is $\frac{5}{12}$.

Using Fractions in Measuring

Fractions can be used to make more careful measurements.

Think about the inch scale on a ruler. Suppose that the spaces between the whole inch marks are left unmarked. With a ruler like this, we can measure only to the nearest inch. When we mark the spaces between inch marks, we can measure to the nearest $\frac{1}{2}$ inch and $\frac{1}{4}$ inch.

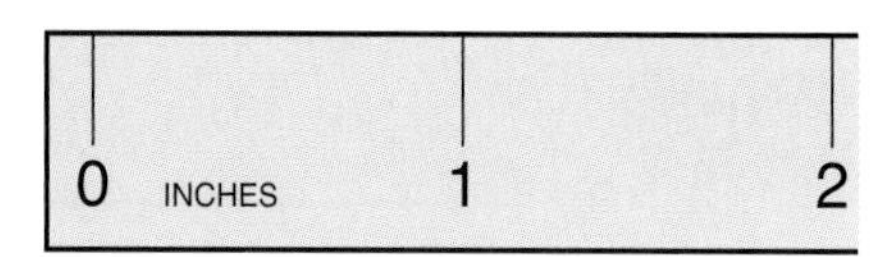

Ruler has inch marks only.
We can measure to nearest inch.

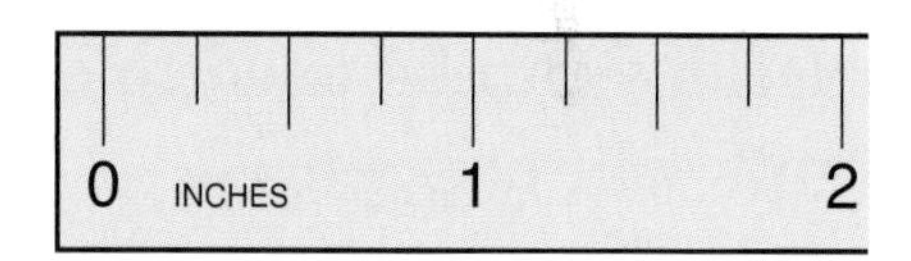

Ruler has $\frac{1}{4}$ and $\frac{1}{2}$ inch marks.
We can measure to nearest $\frac{1}{4}$ inch.

Other measuring tools are marked to show fractions of a unit. Measuring cups usually have marks that show $\frac{1}{4}$-cup, $\frac{1}{3}$-cup, and $\frac{1}{2}$-cup divisions.

Kitchen scales usually have ounce marks that divide the pound marks. 1 pound = 16 ounces. So 1 ounce equals $\frac{1}{16}$ pound.

Be careful. Always say or write the unit that you used in measuring. If you measured an object using an inch scale, then give the length in inches. If you weighed an object to the nearest pound, then give the weight in pounds.

Check Your Understanding

Measure the length of this segment. ______________

1. To the nearest inch

2. To the nearest $\frac{1}{2}$ inch

3. To the nearest $\frac{1}{4}$ inch

Check your answers on page 336.

Other Uses of Fractions

Fractions can name points on a number line that are "in between" the points that are already named.

Examples

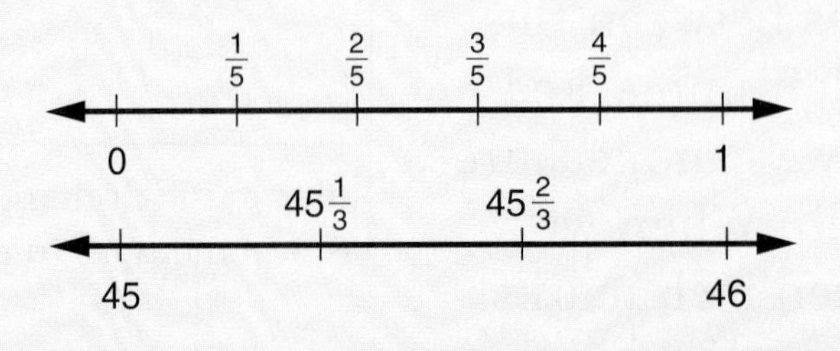

Fractions can be used to describe the chances that events will happen.

Example This spinner has $\frac{1}{3}$ of the circle colored green. And $\frac{2}{3}$ of the circle is not colored.

When we spin this spinner, there is a $\frac{1}{3}$ chance it will land on green.

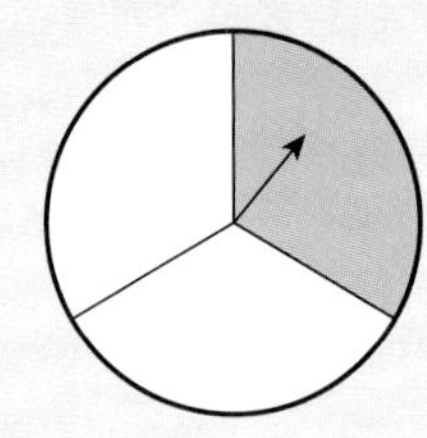

If we spin it many times, it will land on green about $\frac{1}{3}$ of the time.

Fractions can be used to compare two numbers.

Example Kay has \$60. Faith has \$30. Alex has \$10. Compare these amounts.

Faith has $\frac{1}{2}$ as much money as Kay.

Alex has $\frac{1}{3}$ as much money as Faith.

Alex has $\frac{1}{6}$ as much money as Kay.

Equivalent Fractions

Different fractions that name the same amount are called **equivalent fractions.**

Equivalent fractions are equal because they name the same number.

Two circles are shown below. The top half of each circle is shaded. The circles are the same size. The first circle is divided into 2 equal parts. The second is divided into 6 equal parts.

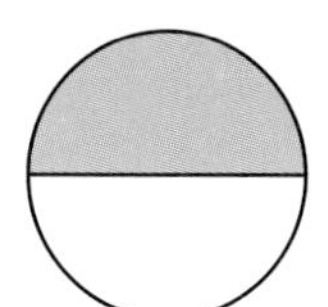

2 equal parts
1 part is shaded
$\frac{1}{2}$ circle is shaded

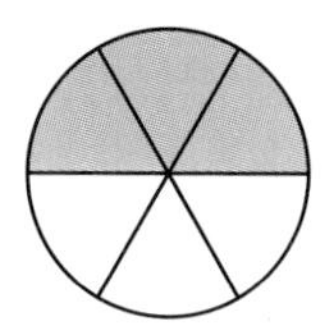

6 equal parts
3 parts are shaded
$\frac{3}{6}$ circle is shaded

The shaded amounts are the same. So $\frac{3}{6}$ of the circle is the same as $\frac{1}{2}$ of the circle.

The fractions $\frac{3}{6}$ and $\frac{1}{2}$ are equivalent fractions.

We write $\frac{3}{6} = \frac{1}{2}$.

Example Eight children go to a party. Two are girls. Six are boys.

$\frac{1}{4}$ of the children are girls

$\frac{2}{8}$ of the children are girls

$\frac{1}{4}$ of the children is the same as $\frac{2}{8}$ of the children.

The fractions $\frac{1}{4}$ and $\frac{2}{8}$ are equivalent fractions.

We write $\frac{1}{4} = \frac{2}{8}$.

Example A rectangle can be used to show fractions that are equivalent to $\frac{3}{4}$.

A rectangle is divided into quarters.
3 quarters are shaded.

So $\frac{3}{4}$ of the rectangle is shaded.

Each quarter is divided into 2 equal parts.
There are 8 equal parts. 6 parts are shaded.

So $\frac{6}{8}$ of the rectangle is shaded.

Each quarter is divided into 3 equal parts.
There are 12 equal parts. 9 parts are shaded.

So $\frac{9}{12}$ of the rectangle is shaded.

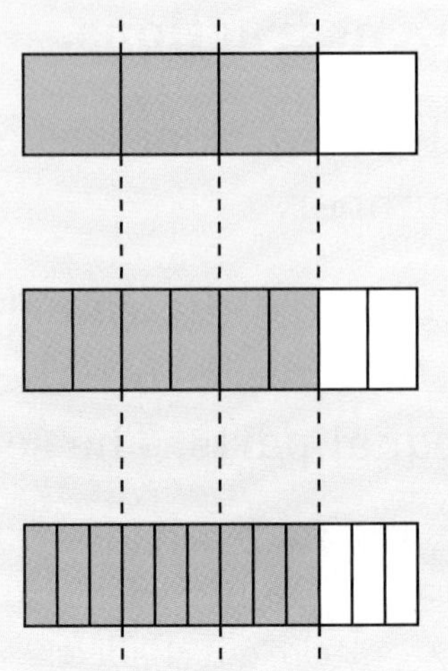

The fractions $\frac{3}{4}$, $\frac{6}{8}$, and $\frac{9}{12}$ all name the same shaded amount. They are all equivalent fractions.

Example Equivalent fractions can be used to read the inch scale on a ruler.

The tip of the nail falls at the $\frac{3}{4}$-inch mark.

The length can be read as $\frac{3}{4}$ in. or $\frac{6}{8}$ in. or $\frac{12}{16}$ in.

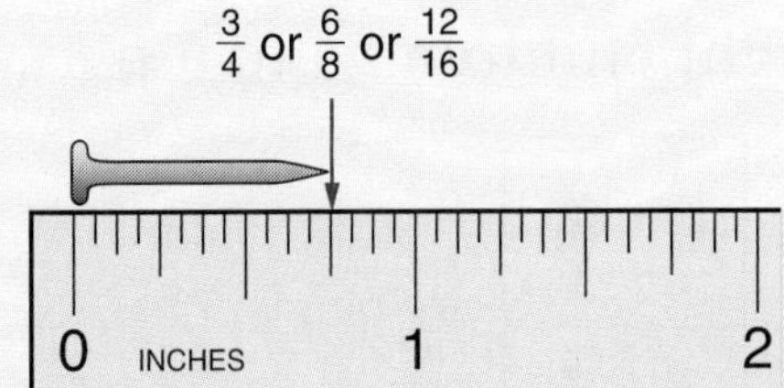

$\frac{3}{4}$, $\frac{6}{8}$, and $\frac{12}{16}$ all name the same mark on the ruler.

They are equivalent fractions.

There are two shortcut rules for finding **equivalent fractions.**

Method 1: Write the fraction. Then multiply both the numerator and the denominator by the same number.

Example Find fractions that are equivalent to $\frac{1}{3}$.

$\frac{1 \times \mathbf{2}}{3 \times \mathbf{2}} = \frac{2}{6}$ $\frac{1 \times \mathbf{4}}{3 \times \mathbf{4}} = \frac{4}{12}$ $\frac{1 \times \mathbf{10}}{3 \times \mathbf{10}} = \frac{10}{30}$

$\frac{1}{3}$, $\frac{2}{6}$, $\frac{4}{12}$, and $\frac{10}{30}$ are all equivalent fractions.

These fractions are all equal:

$\frac{1}{3} = \frac{2}{6}$ $\frac{1}{3} = \frac{4}{12}$ $\frac{1}{3} = \frac{10}{30}$ $\frac{2}{6} = \frac{4}{12}$ $\frac{2}{6} = \frac{10}{30}$ $\frac{4}{12} = \frac{10}{30}$

Method 2: Write the fraction. Then divide both the numerator and the denominator by the same number.

Example Find fractions that are equivalent to $\frac{12}{24}$.

$\frac{12 \div \mathbf{6}}{24 \div \mathbf{6}} = \frac{2}{4}$ $\frac{12 \div \mathbf{4}}{24 \div \mathbf{4}} = \frac{3}{6}$ $\frac{12 \div \mathbf{12}}{24 \div \mathbf{12}} = \frac{1}{2}$

$\frac{12}{24}$, $\frac{2}{4}$, $\frac{3}{6}$, and $\frac{1}{2}$ are all equivalent fractions.

These fractions are all equal:

$\frac{12}{24} = \frac{2}{4}$ $\frac{12}{24} = \frac{3}{6}$ $\frac{12}{24} = \frac{1}{2}$ $\frac{2}{4} = \frac{3}{6}$ $\frac{2}{4} = \frac{1}{2}$ $\frac{3}{6} = \frac{1}{2}$

Check Your Understanding

Write 5 fractions that are equivalent to $\frac{20}{30}$.

Check your answers on page 336.

Table of Equivalent Fractions

The table below lists equivalent fractions. All of the fractions in a row name the same number.

For example, all the fractions in the last row are names for the number $\frac{5}{6}$. So, all of these fractions are equal:

$\frac{5}{6} = \frac{10}{12}$, $\frac{5}{6} = \frac{15}{18}$, $\frac{10}{12} = \frac{15}{18}$, and so on.

Simplest Name	Equivalent Fraction Names								
0 (zero)	$\frac{0}{1}$	$\frac{0}{2}$	$\frac{0}{3}$	$\frac{0}{4}$	$\frac{0}{5}$	$\frac{0}{6}$	$\frac{0}{7}$	$\frac{0}{8}$	$\frac{0}{9}$
1 (one)	$\frac{1}{1}$	$\frac{2}{2}$	$\frac{3}{3}$	$\frac{4}{4}$	$\frac{5}{5}$	$\frac{6}{6}$	$\frac{7}{7}$	$\frac{8}{8}$	$\frac{9}{9}$
$\frac{1}{2}$	$\frac{2}{4}$	$\frac{3}{6}$	$\frac{4}{8}$	$\frac{5}{10}$	$\frac{6}{12}$	$\frac{7}{14}$	$\frac{8}{16}$	$\frac{9}{18}$	$\frac{10}{20}$
$\frac{1}{3}$	$\frac{2}{6}$	$\frac{3}{9}$	$\frac{4}{12}$	$\frac{5}{15}$	$\frac{6}{18}$	$\frac{7}{21}$	$\frac{8}{24}$	$\frac{9}{27}$	$\frac{10}{30}$
$\frac{2}{3}$	$\frac{4}{6}$	$\frac{6}{9}$	$\frac{8}{12}$	$\frac{10}{15}$	$\frac{12}{18}$	$\frac{14}{21}$	$\frac{16}{24}$	$\frac{18}{27}$	$\frac{20}{30}$
$\frac{1}{4}$	$\frac{2}{8}$	$\frac{3}{12}$	$\frac{4}{16}$	$\frac{5}{20}$	$\frac{6}{24}$	$\frac{7}{28}$	$\frac{8}{32}$	$\frac{9}{36}$	$\frac{10}{40}$
$\frac{3}{4}$	$\frac{6}{8}$	$\frac{9}{12}$	$\frac{12}{16}$	$\frac{15}{20}$	$\frac{18}{24}$	$\frac{21}{28}$	$\frac{24}{32}$	$\frac{27}{36}$	$\frac{30}{40}$
$\frac{1}{5}$	$\frac{2}{10}$	$\frac{3}{15}$	$\frac{4}{20}$	$\frac{5}{25}$	$\frac{6}{30}$	$\frac{7}{35}$	$\frac{8}{40}$	$\frac{9}{45}$	$\frac{10}{50}$
$\frac{2}{5}$	$\frac{4}{10}$	$\frac{6}{15}$	$\frac{8}{20}$	$\frac{10}{25}$	$\frac{12}{30}$	$\frac{14}{35}$	$\frac{16}{40}$	$\frac{18}{45}$	$\frac{20}{50}$
$\frac{3}{5}$	$\frac{6}{10}$	$\frac{9}{15}$	$\frac{12}{20}$	$\frac{15}{25}$	$\frac{18}{30}$	$\frac{21}{35}$	$\frac{24}{40}$	$\frac{27}{45}$	$\frac{30}{50}$
$\frac{4}{5}$	$\frac{8}{10}$	$\frac{12}{15}$	$\frac{16}{20}$	$\frac{20}{25}$	$\frac{24}{30}$	$\frac{28}{35}$	$\frac{32}{40}$	$\frac{36}{45}$	$\frac{40}{50}$
$\frac{1}{6}$	$\frac{2}{12}$	$\frac{3}{18}$	$\frac{4}{24}$	$\frac{5}{30}$	$\frac{6}{36}$	$\frac{7}{42}$	$\frac{8}{48}$	$\frac{9}{54}$	$\frac{10}{60}$
$\frac{5}{6}$	$\frac{10}{12}$	$\frac{15}{18}$	$\frac{20}{24}$	$\frac{25}{30}$	$\frac{30}{36}$	$\frac{35}{42}$	$\frac{40}{48}$	$\frac{45}{54}$	$\frac{50}{60}$

Did You Know?

The word *fraction* is derived from a Latin word that means "to break." Fractions are sometimes called "broken numbers."

Arab mathematicians began to use the horizontal fraction bar around the year 1200. They were the first to write fractions as we do today.

Comparing Fractions to $\frac{1}{2}$, 0, and 1

Shading on the Fraction Cards makes it clear whether a fraction is less than $\frac{1}{2}$, greater than $\frac{1}{2}$, or equal to $\frac{1}{2}$.

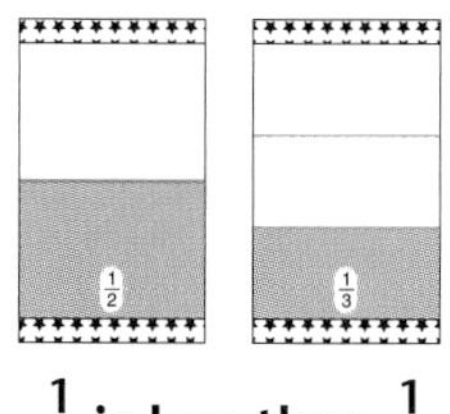

$\frac{1}{3}$ is less than $\frac{1}{2}$

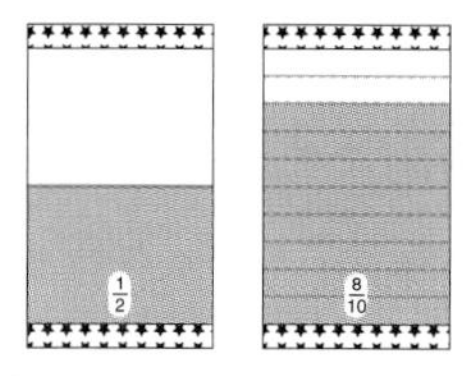

$\frac{8}{10}$ is greater than $\frac{1}{2}$

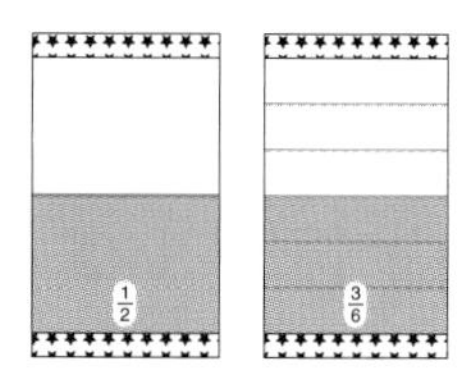

$\frac{3}{6}$ is equal to $\frac{1}{2}$

You can also compare a fraction to $\frac{1}{2}$ by looking at the numerator and denominator.

- If the numerator is less than half of the denominator, the fraction is less than $\frac{1}{2}$.

  For example, in $\frac{1}{3}$, 1 is less than half of 3.

- If the numerator is more than half of the denominator, the fraction is more than $\frac{1}{2}$.

  For example, in $\frac{8}{10}$, 8 is more than half of 10.

- If the numerator is exactly half of the denominator, the fraction is equal to $\frac{1}{2}$.

  For example, in $\frac{3}{6}$, 3 is half of 6.

You can use the greater-than symbol ($>$) or the less-than symbol ($<$) when comparing fractions.

Examples

$\frac{8}{10} > \frac{1}{2}$ means that $\frac{8}{10}$ is greater than $\frac{1}{2}$.

$\frac{1}{2} < \frac{8}{10}$ means that $\frac{1}{2}$ is less than $\frac{8}{10}$.

$\frac{4}{6} > \frac{1}{2}$ means that $\frac{4}{6}$ is greater than $\frac{1}{2}$.

Shading on the Fraction Cards also makes it clear whether a fraction is close to 0 or close to 1.

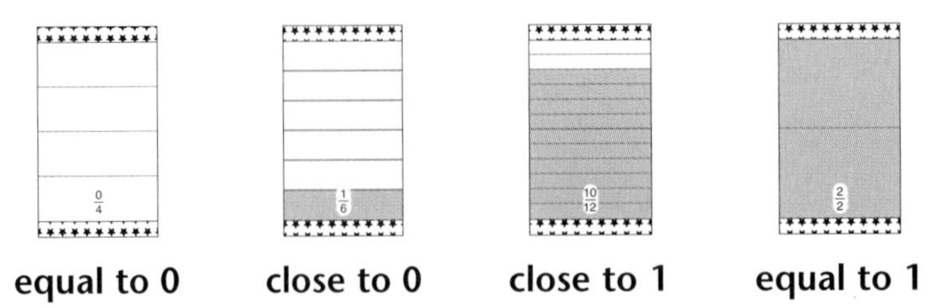

equal to 0 close to 0 close to 1 equal to 1

You can also compare a fraction to 0 and 1 by looking at the numerator and denominator.

- If the numerator of the fraction is small compared to the denominator, the fraction is close to 0.

  For example, in $\frac{20}{1{,}000}$, 20 is small compared to 1,000.

  So $\frac{20}{1{,}000}$ is close to 0.

- If the numerator of the fraction is close to the denominator, the fraction is close to 1.

  For example, in $\frac{20}{21}$, 20 is close to 21. So $\frac{20}{21}$ is close to 1.

Check Your Understanding

Compare each fraction to $\frac{1}{2}$. Use $<$, $>$, or $=$.

1. $\frac{1}{2} \square \frac{3}{8}$ **2.** $\frac{1}{2} \square \frac{6}{10}$ **3.** $\frac{2}{4} \square \frac{1}{2}$

4. $\frac{3}{10} \square \frac{1}{2}$ **5.** $\frac{1}{2} \square \frac{30}{60}$ **6.** $\frac{1}{3} \square \frac{1}{2}$

Use the Fraction Cards to help you. Write *close to 1* or *close to 0*.

7. $\frac{2}{10}$ **8.** $\frac{7}{8}$ **9.** $\frac{4}{5}$ **10.** $\frac{2}{12}$ **11.** $\frac{0}{8}$ **12.** $\frac{46}{50}$

Check your answers on page 336.

Decimals

The numbers 0.3, 7.4, 0.46, and 23.456 are all **decimals.** Decimals are another way to write fractions.

Decimals can be used in the same ways that fractions are used. You may use decimals:

- To name a part of a whole thing or a part of a collection.
- To make more careful measurements of length, weight, time, and so on.

Money amounts are decimals with a dollar sign in front of the number. In the amount $3.42, the 3 names whole dollars and the 42 names part of a dollar. The dot is called a **decimal point.**

Decimal names follow a simple pattern:

- Decimals that have **1 digit** after the decimal point are **"tenths."** For example, 0.3 is another way to write the fraction $\frac{3}{10}$ (3 tenths).
- Decimals that have **2 digits** after the decimal point are **"hundredths."** For example, 0.25 is another way to write the fraction $\frac{25}{100}$ (25 hundredths).

Pattern for Tenths

Fraction Name	Read the Number	Decimal Name
$\frac{1}{10}$	1 tenth	0.1
$\frac{2}{10}$	2 tenths	0.2
$\frac{5}{10}$	5 tenths	0.5
$\frac{8}{10}$	8 tenths	0.8

Pattern for Hundredths

Fraction Name	Read the Number	Decimal Name
$\frac{1}{100}$	1 hundredth	0.01
$\frac{5}{100}$	5 hundredths	0.05
$\frac{13}{100}$	13 hundredths	0.13
$\frac{20}{100}$	20 hundredths	0.20

The pattern continues for thousandths. The fraction $\frac{1}{1,000}$ is written as the decimal 0.001. The fraction $\frac{114}{1,000}$ is written as the decimal 0.114. The fraction name and the decimal name are both read as "114 thousandths." A decimal name that has **3 digits** after the decimal is read as **"thousandths."**

Examples Read the decimals 0.15, 0.015, 98.7, and 98.107.

0.15 is read as "15 hundredths" and is another name for $\frac{15}{100}$.

0.015 is read as "15 thousandths" and is another name for $\frac{15}{1,000}$.

98.7 is read as "98 and 7 tenths" and is another name for $98\frac{7}{10}$.

98.107 is read as "98 and 107 thousandths" and is another name for $98\frac{107}{1,000}$.

Example How much of the square is shaded? Give the fraction and decimal names.

The square is divided into 100 equal parts.

Each part is $\frac{1}{100}$ of the square.

The decimal name for $\frac{1}{100}$ is 0.01 (1 hundredth).

42 squares are shaded.

So $\frac{42}{100}$ of the square is shaded. The decimal name for $\frac{42}{100}$ is 0.42 (42 hundredths).

Check Your Understanding

How much of each square is shaded?

Give the fraction and decimal names.

1.

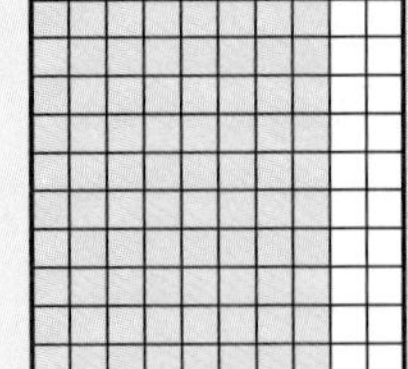

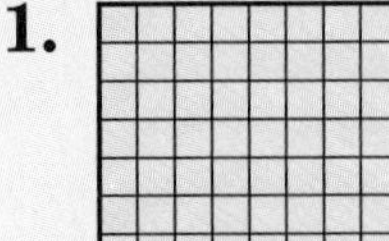

2.

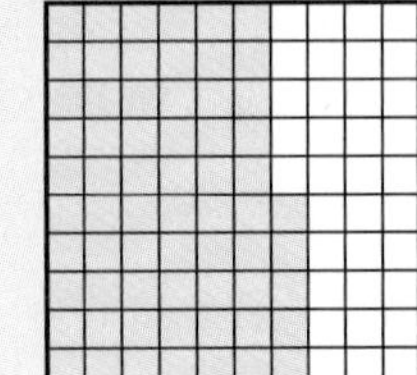

Check your answers on page 336.

Place Value for Decimals

When we write a money amount like $6.23, the number is a decimal. The place that each digit has in the number is very important.

dollars		dimes	pennies
6	.	2	3

Did You Know?

Decimals were invented by the Dutch scientist Simon Stevin, in 1585.

In England, 3.42 is written as 3·42. In France, 3.42 is written as 3,42.

The decimal point separates dollars from cents.

The 6 is worth 6 dollars.

The 2 is worth 20 cents, or 2 dimes, or $\frac{2}{10}$ of a dollar.

The 3 is worth 3 cents, or 3 pennies, or $\frac{3}{100}$ of a dollar.

We can use a **place-value chart** to show how much each digit in a decimal is worth.
The **place** for a digit is its position in the number.
The **value** of a digit is how much it is worth.

Example The number 3.456 is shown in a place-value chart below.

1s ones place		**0.1s** tenths place	**0.01s** hundredths place	**0.001s** thousandths place
3	.	4	5	6

The 3 in the ones place is worth 3 (3 ones).
The 4 in the tenths place is worth 0.4 (4 tenths).
The 5 in the hundredths place is worth 0.05 (5 hundredths).
The 6 in the thousandths place is worth 0.006 (6 thousandths).

3.456 is read "3 and 456 thousandths." The decimal point is read as "and."

Comparing Decimals

A place-value chart can help us to compare two decimal numbers.

Example Compare the numbers 9.235 and 9.253.

1s ones place		**0.1s** tenths place	**0.01s** hundredths place	**0.001s** thousandths place
9	.	2	3	5
9	.	2	5	3

Start on the left side.
The ones digits *are* the same. They are both worth 9.
The tenths digits *are* the same. They are worth $\frac{2}{10}$, or 0.2.
The hundredths digits are *not* the same. The 3 is worth $\frac{3}{100}$, or 0.03, and the 5 is worth $\frac{5}{100}$, or 0.05. The 5 is worth more.

So 9.253 is larger than 9.235. $9.253 > 9.235$

Example Which number is larger, 3.4 or 3.40?

The ones digits are the same. The tenths digits are the same.
The hundredths place in the number 3.40 has a 0. It is worth $\frac{0}{100}$, which is 0.
So the hundredths place in 3.40 has no value.

3.4 and 3.40 are equal. $3.4 = 3.40$

If you write a zero at the end of a decimal number that has a decimal point, the value of the decimal will not change.

Check Your Understanding

Which number is larger?

1. 4.36 or 4.6 **2.** 0.6 or 0.572 **3.** 1.4 or 1.04

Check your answers on page 336.

Factors of a Number and Prime Numbers

When two numbers are multiplied, the number answer is called the **product.** The two numbers that are multiplied are called **factors** of the product.

Example $3 \times 5 = 15$. 15 is the product of 3 and 5. 3 is a factor of 15. 5 is another factor of 15.

When you are asked to find the factors of a counting number, those factors must be counting numbers.

Example Find all the factors of 24.

Ways to Write 24	Factors of 24
$1 \times 24 = 24$	1 and 24
$2 \times 12 = 24$	2 and 12
$3 \times 8 = 24$	3 and 8
$4 \times 6 = 24$	4 and 6

The numbers 1, 2, 3, 4, 6, 8, 12, and 24 are all factors of 24. And they are the only factors. The only ways to multiply two counting numbers and get 24 are shown above.

Example Find all the factors of 13. There is only one way to multiply two counting numbers and get 13. $1 \times 13 = 13$. So 1 and 13 are both factors of 13. And they are the only factors of 13.

A counting number that has exactly two different factors is called a **prime number.**
A counting number that has 3 or more different factors is called a **composite number.**

Even and Odd Numbers

- A counting number is an **even number** if 2 is one of its factors.
- A counting number is an **odd number** if it is not an even number.

Facts about the Numbers 1 to 20

Number	Factors	Prime or Composite	Even or Odd
1	1	neither	odd
2	1 and 2	prime	even
3	1 and 3	prime	odd
4	1, 2, and 4	composite	even
5	1 and 5	prime	odd
6	1, 2, 3, and 6	composite	even
7	1 and 7	prime	odd
8	1, 2, 4, and 8	composite	even
9	1, 3, and 9	composite	odd
10	1, 2, 5, and 10	composite	even
11	1 and 11	prime	odd
12	1, 2, 3, 4, 6, and 12	composite	even
13	1 and 13	prime	odd
14	1, 2, 7, and 14	composite	even
15	1, 3, 5, and 15	composite	odd
16	1, 2, 4, 8, and 16	composite	even
17	1 and 17	prime	odd
18	1, 2, 3, 6, 9, and 18	composite	even
19	1 and 19	prime	odd
20	1, 2, 4, 5, 10, and 20	composite	even

The number 1 has only one factor. The only way to multiply two counting numbers and get 1 is to write $1 \times 1 = 1$. So 1 is the only factor of the number 1.

Prime and composite numbers have at least 2 different factors. So the number 1 is not prime, and it is not composite.

Negative Numbers

Positive numbers are numbers that are greater than 0. **Negative numbers** are numbers that are less than 0.

The numbers –1, –2, –3, $-\frac{1}{2}$, and –31.6 are all negative numbers. The number –2 is read "negative 2."

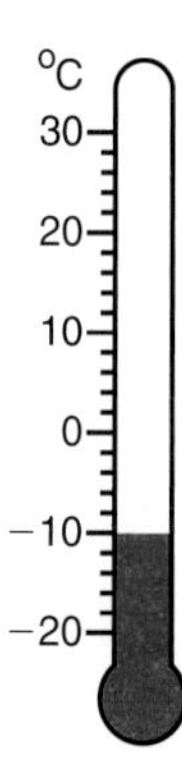

Example The scale on a thermometer often shows both positive and negative numbers. A temperature of –10°C is read "10 degrees below zero."

Example Many number lines include negative numbers.

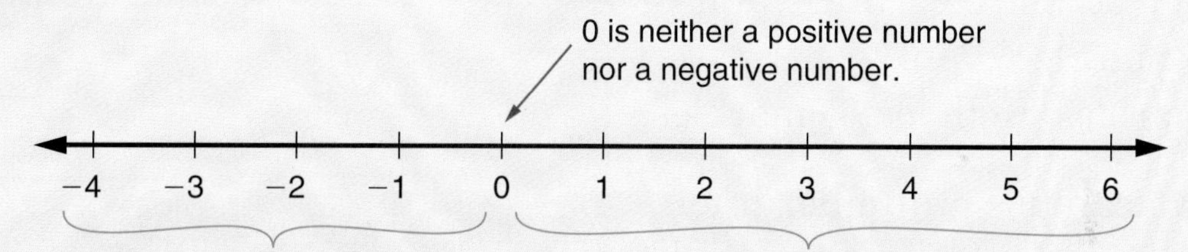

Examples The tables on page 240 show distances above and below sea level for different places. Negative numbers are used for places below sea level.

Death Valley, California is 282 feet below sea level (written –282 ft).

New Orleans, Louisiana is 8 feet below sea level (written –8 ft).

A number grid is often shown starting at the number 0. Think of counting backward by 1s on a number grid. Do not stop at 0. The number grid will now include negative numbers.

−19	−18	−17	−16	−15	−14	−13	−12	−11	−10
−9	−8	−7	−6	−5	−4	−3	−2	−1	0
1	2	3	4	5	6	7	8	9	10

Negative numbers can be used to show changes. You might use negative numbers to show yards lost in a football game or to show how much weight someone lost.

Example Jack, Paul, and Kelli played marbles. Jack won 7 marbles. Paul lost 2 marbles. Kelli lost 5 marbles.

We can use a number line to show both the gains and the losses. The number line must include both positive and negative numbers.

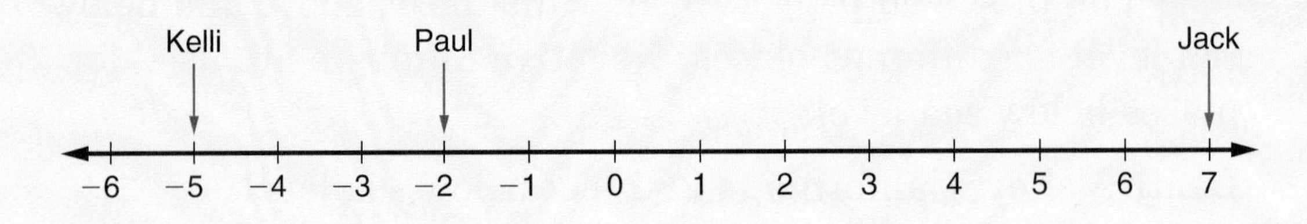

Very Large and Very Small Numbers

Earth weighs about 1,175,800,000,000,000,000,000,000 pounds, give or take a few billion. Here is how you say the weight: one *septillion,* one hundred seventy-five *sextillion,* eight hundred *quintillion* pounds. A *septillion* is written with a 1 followed by twenty-four 0s.

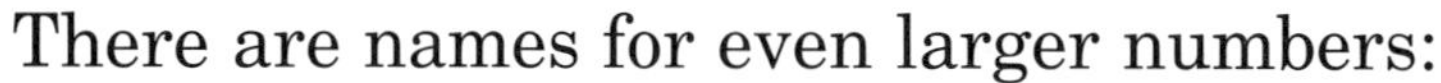

There are names for even larger numbers:

novemdecillion: a 1 followed by sixty 0s
googol: a 1 followed by one hundred 0s

How about this for a tongue twister:

A quintoquadagintillion is 1 followed by one hundred thirty-eight 0s!

Large numbers of things don't always take up a lot of space. The blood of a human being contains tiny cells called *red corpuscles.* There are about 5 million of these corpuscles in 1 cubic millimeter.

Try to imagine 5 million tiny cells in a cubic millimeter! A cubic millimeter takes up very little space. It is a cube whose six faces all look like this red square ▪. Each side of the red square is 1 millimeter long.

Did You Know?

Google™ is a search engine that you may use to search for information on the Internet. Google organizes a huge amount of information. This is why a form of the word *googol* was used to name the search engine.

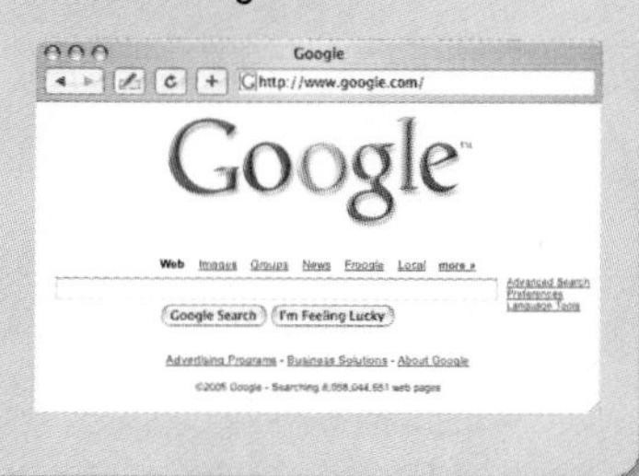

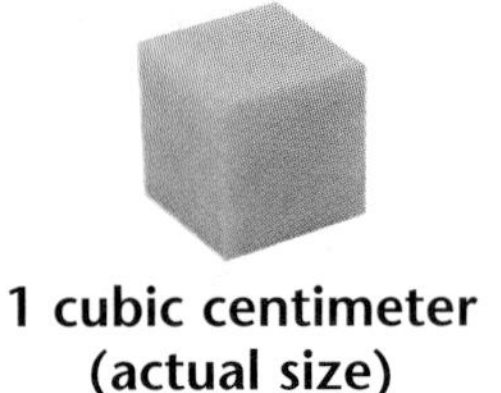
1 cubic centimeter (actual size)

1,000 cubic millimeters fit into 1 cubic centimeter, so there are about 5 billion red corpuscles in 1 cubic centimeter. That's a 5 followed by nine 0s, or 5,000,000,000.

There are almost 1,000 cubic centimeters in 1 quart. The body of a healthy adult contains about 5 quarts of blood. So an adult has about 25 trillion red corpuscles. That's a 25 followed by twelve 0s, or 25,000,000,000,000.

Do you see why we need to have names for very large numbers?

Very small numbers may be even harder to imagine than very large numbers. A beam of light would cover the distance from one side of the classroom to the other in about 0.000000020, or 20 billionths, of a second. Another name for one billionth of a second is a nanosecond. Think of it. Light travels about 10 million times as fast as a car that is going 70 miles per hour.

Did You Know?

Light travels about 186,000 miles per second.

Sound travels about 1,115 feet per second. That is about $\frac{1}{5}$ mile per second.

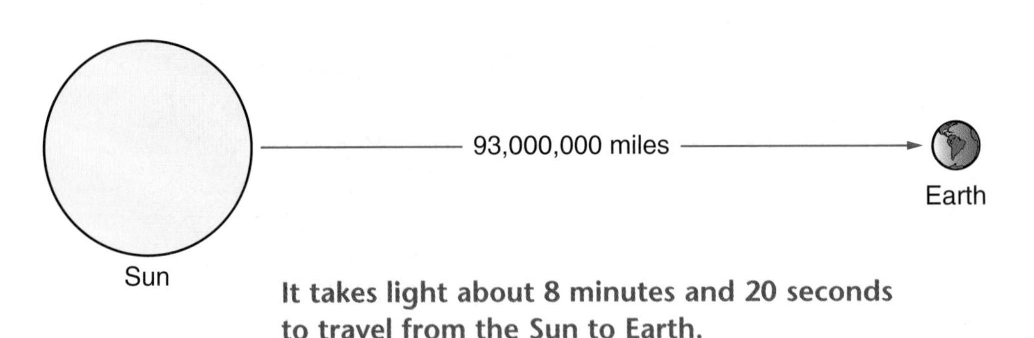

It takes light about 8 minutes and 20 seconds to travel from the Sun to Earth.

A History of Counting and Calculating

Mathematics ... Every Day

In the thousands of years since people began to count, we have developed many different number systems and tools for counting and calculating.

We learn to count on our fingers at a very young age. ➤

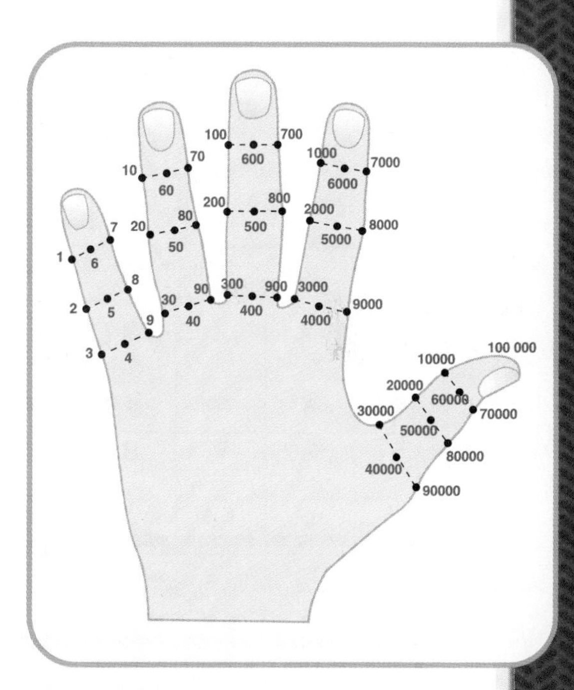

Many of us use our hands as simple calculators when we start learning to add and subtract. As we memorize math facts, we rely on our fingers less and less.

In China, a method has been developed to count up to 100 thousand using one hand and up to 10 billion using two hands. Do you see the pattern?

Writing Numbers

Different societies have used different types of systems for writing numbers.

This ancient Babylonian clay tablet shows a math problem.

1	11	21	31	41	51
2	12	22	32	42	52
3	13	23	33	43	53
4	14	24	34	44	54
5	15	25	35	45	55
6	16	26	36	46	56
7	17	27	37	47	57
8	18	28	38	48	58
9	19	29	39	49	59
10	20	30	40	50	

The Babylonians of the Middle East used 60 as the base for their number system. We still use base-60 today in measuring time, angles, and latitude and longitude.

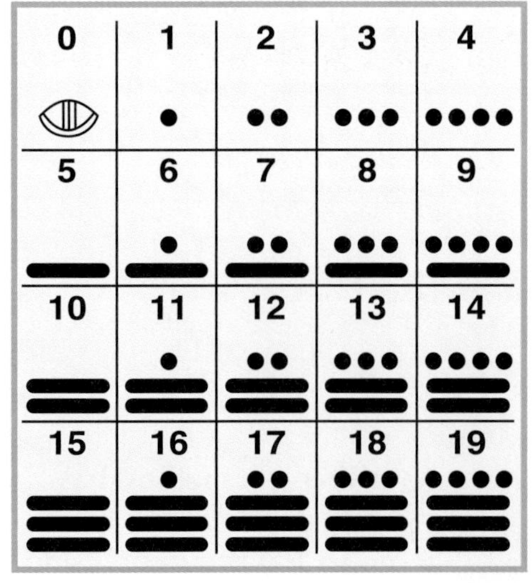

The ancient Maya of Central America chose 20 as the base for their number system, perhaps because they counted using their fingers and toes.

◄ Roman numerals came into use more than 2,000 years ago. They are still used today for formal purposes. For example, people use Roman numerals on buildings and clocks, and for sporting events.

This is a gold medal from the 27th modern Olympic Games in Sydney, Australia. ►

The Western world today uses Arabic numerals. This base-10 system requires ten different numerals. Arabic numerals, like Mayan numerals, have a symbol for zero. Roman numerals do not. ►

Tools to Count and Calculate

The tools people use to count and calculate have improved with advances in technology.

Archaeologists have discovered 34,000 year-old animal bones with notches made by people. These notches may have been tallies of hunting successes. ➤

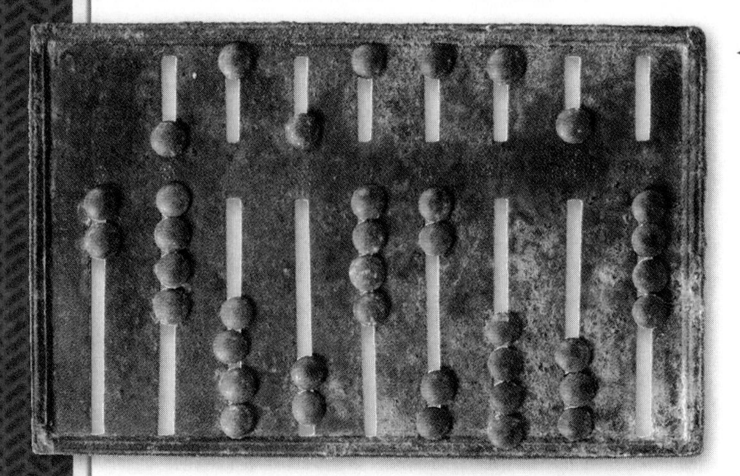

➤ The abacus is a powerful calculating tool that can be used to add, subtract, multiply, and divide. The Roman abacus was used in Italy as far back as the second century A.D.

The table abacus was used in Europe until the late 1600s. Coin-like discs were moved on lines that were drawn on a table or cloth. ➤

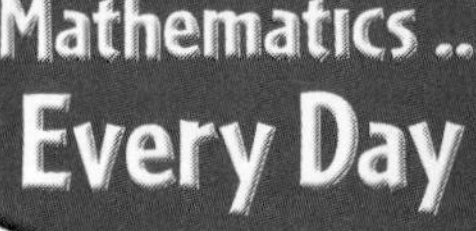

Lower Deck

➤ The bead abacus was created in China in about 1200 A.D. Beads are counted by moving them to the beam that separates the upper and lower decks. After 5 beads are counted in the lower deck, the result is "carried" to the upper deck.

The abacus is still used for counting and calculating in many parts of the world today. This abacus repairman in modern day Hong Kong has no shortage of work. ➤

➤ This woman in a small shop in Volgograd, Russia uses an abacus to calculate what her customers owe.

Electronic Calculators

The abacus is being replaced by the calculator in most parts of the world. The calculator can do things that are not possible on an abacus, such as extract square roots easily.

Handheld calculators were first sold in 1971. They are smaller and more portable than the abacus. One or more silicon chips act as the "brains" of the calculator. ➤

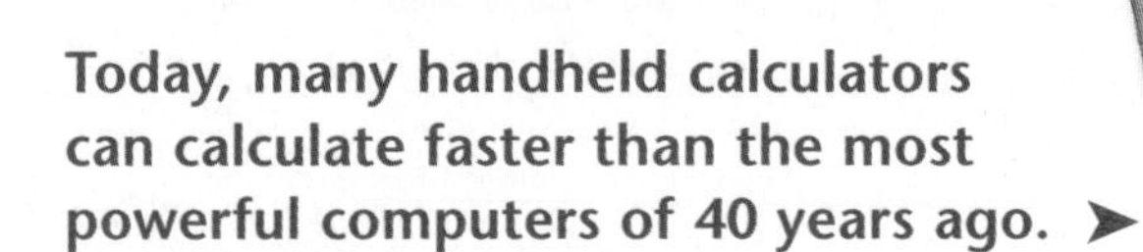

Today, many handheld calculators can calculate faster than the most powerful computers of 40 years ago. ➤

As time goes on, humans will invent more powerful ways to calculate. But no matter how advanced technology becomes, people will continue to use their fingers and brains for counting and calculating.

What methods do you use to count, add, subtract, multiply, and divide?

Operations and Computation

Basic Facts for Addition and Subtraction

Reading is easier when you know the words by sight. In mathematics, solving problems is easier when you know the basic number facts. Here are some examples of basic addition and subtraction facts.

Basic addition facts:

$6 + 4 = 10,\ 0 + 7 = 7,\ 3 + 5 = 8,\ 9 + 9 = 18$

Basic subtraction facts:

$10 - 6 = 4,\ 7 - 7 = 0,\ 8 - 5 = 3,\ 18 - 9 = 9$

The **facts table** shown below is a chart with rows and columns. It can be used to show *all* of the basic addition and subtraction facts.

Addition/Subtraction Facts Table

+,−	0	1	2	3	4	5	6	7	8	9
0	0	1	2	3	4	5	6	7	8	9
1	1	2	3	4	5	6	7	8	9	10
2	2	3	4	5	6	7	8	9	10	11
3	3	4	5	6	7	8	9	10	11	12
4	4	5	6	7	8	9	10	11	12	13
5	5	6	7	8	9	10	11	12	13	14
6	6	7	8	9	10	11	12	13	14	15
7	7	8	9	10	11	12	13	14	15	16
8	8	9	10	11	12	13	14	15	16	17
9	9	10	11	12	13	14	15	16	17	18

Did You Know?

In 1489, Johann Widmann wrote the first book that used the + and − signs. Traders and shopkeepers had used both of the signs long before this. They used + to show they had too much of something. And they used − to show they had too little of something.

The facts table can be used to find *all* of the basic addition and subtraction facts.

Example Which addition facts and subtraction facts can you find using the 4-row and the 6-column?

Go across the 4-row while you go down the 6-column. This row and column meet at a square that shows the number 10.

The numbers 4, 6, and 10 can be used to write two addition facts and two subtraction facts:

$4 + 6 = 10$ $10 - 4 = 6$
$6 + 4 = 10$ $10 - 6 = 4$

6-column

4-row

+,−	0	1	2	3	4	5	6	7	8	9
0	0	1	2	3	4	5	6	7	8	9
1	1	2	3	4	5	6	7	8	9	10
2	2	3	4	5	6	7	8	9	10	11
3	3	4	5	6	7	8	9	10	11	12
4	4	5	6	7	8	9	10	11	12	13
5	5	6	7	8	9	10	11	12	13	14
6	6	7	8	9	10	11	12	13	14	15
7	7	8	9	10	11	12	13	14	15	16
8	8	9	10	11	12	13	14	15	16	17
9	9	10	11	12	13	14	15	16	17	18

Check Your Understanding

Use the facts table above. Write the addition and subtraction facts you can find.

1. Use the 8-row and the 9-column.
2. Use the 7-row and the 5-column.
3. Use the 3-row and the 8-column.
4. Use the 9-row and the 9-column.

Check your answers on page 337.

Basic Facts for Multiplication and Division

Solving problems is easier when you know the basic number facts. Here are some examples of basic multiplication and division facts:

Basic multiplication facts:

$6 \times 4 = 24,\ 10 \times 7 = 70,\ 1 \times 8 = 8,\ 3 \times 9 = 27$

Basic division facts:

$24 \div 6 = 4,\ 70 \div 10 = 7,\ 8 \div 1 = 8,\ 27 \div 3 = 9$

The **facts table** shown below is a chart with rows and columns. It can be used to find *all* of the basic multiplication and division facts.

Multiplication/Division Facts Table

×,÷	1	2	3	4	5	6	7	8	9	10
1	1	2	3	4	5	6	7	8	9	10
2	2	4	6	8	10	12	14	16	18	20
3	3	6	9	12	15	18	21	24	27	30
4	4	8	12	16	20	24	28	32	36	40
5	5	10	15	20	25	30	35	40	45	50
6	6	12	18	24	30	36	42	48	54	60
7	7	14	21	28	35	42	49	56	63	70
8	8	16	24	32	40	48	56	64	72	80
9	9	18	27	36	45	54	63	72	81	90
10	10	20	30	40	50	60	70	80	90	100

The facts table can be used to find *all* of the basic multiplication and division facts.

Example Which multiplication facts and division facts can you write using the 4-row and the 6-column?

Go across the 4-row to the 6-column. This row and column meet at a square that shows the number 24.

The numbers 4, 6, and 24 can be used to write two multiplication facts and two division facts:

$4 \times 6 = 24$ $24 \div 4 = 6$

$6 \times 4 = 24$ $24 \div 6 = 4$

6-column

4-row

×,÷	1	2	3	4	5	6	7	8	9	10
1	1	2	3	4	5	6	7	8	9	10
2	2	4	6	8	10	12	14	16	18	20
3	3	6	9	12	15	18	21	24	27	30
4	4	8	12	16	20	24	28	32	36	40
5	5	10	15	20	25	30	35	40	45	50
6	6	12	18	24	30	36	42	48	54	60
7	7	14	21	28	35	42	49	56	63	70
8	8	16	24	32	40	48	56	64	72	80
9	9	18	27	36	45	54	63	72	81	90
10	10	20	30	40	50	60	70	80	90	100

Check Your Understanding

Use the facts table above. Write the multiplication and division facts you can find.

1. Use the 8-row and the 9-column.
2. Use the 7-row and the 5-column.
3. Use the 3-row and the 8-column.
4. Use the 9-row and the 9-column.

Check your answers on page 337.

Fact Triangles and Fact Families

Fact Triangles are tools that can help you memorize the basic facts. One set of Fact Triangles is used to practice addition and subtraction facts. A second set of Fact Triangles is used to practice multiplication and division facts.

Here is a Fact Triangle card. The "+,−" printed on the card means that it is used to practice addition and subtraction facts. The number in the • corner is the sum of the other two numbers.

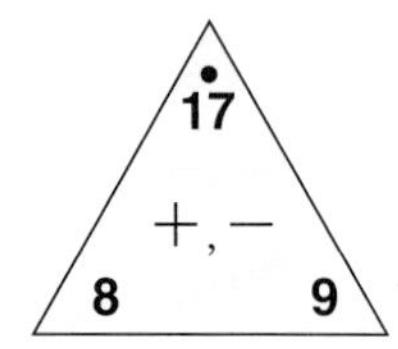

Fact family for this Fact Triangle

$8 + 9 = 17$ $\quad$ $17 - 8 = 9$

$9 + 8 = 17$ $\quad$ $17 - 9 = 8$

A Fact Triangle shows basic facts for the numbers printed on the card. These facts are called a **fact family.**

Work with a partner when you use Fact Triangles to practice facts. One partner covers one of the three corners with a finger. The other partner gives an addition or subtraction fact.

Example Here are the ways to use the Fact Triangle shown above.

Marla covers 17. Alice says "8 + 9 equals 17" or "9 + 8 equals 17."

Marla covers 9. Alice says "17 − 8 equals 9."

Marla covers 8. Alice says "17 − 9 equals 8."

Here is a Fact Triangle card. The "×, ÷" printed on the card means that it is used to practice multiplication and division facts. The number in the • corner is the product of the other two numbers.

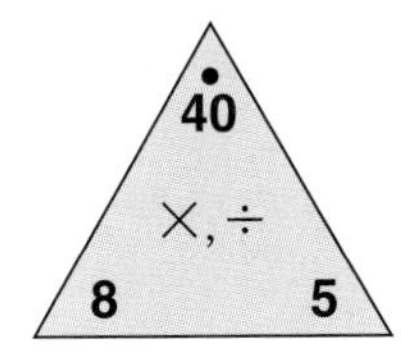

Fact family for this Fact Triangle

$5 \times 8 = 40$ $40 \div 5 = 8$

$8 \times 5 = 40$ $40 \div 8 = 5$

Check Your Understanding

1. Write the fact family for each of these Fact Triangles.

a. 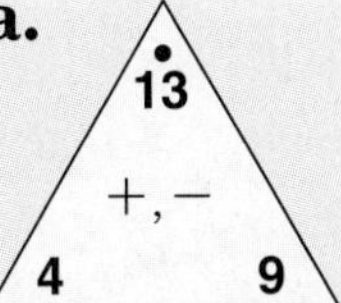

b. 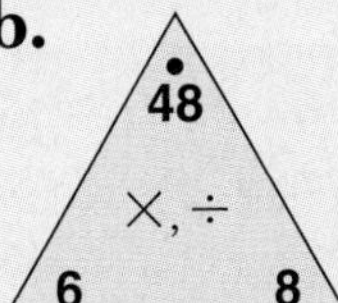

2. Draw a Fact Triangle for each of these fact families. Write in the three numbers for each triangle.

a. $7 + 5 = 12$

$5 + 7 = 12$

$12 - 5 = 7$

$12 - 7 = 5$

b. $6 \times 10 = 60$

$10 \times 6 = 60$

$60 \div 6 = 10$

$60 \div 10 = 6$

Check your answers on page 337.

Shortcuts

Here are some ways to use facts you know to learn new facts. They are called shortcuts.

Plus 0: If 0 is added to a number, the number is not changed.

Examples $6 + 0 = 6$ $0 + 812 = 812$

Minus 0: If 0 is subtracted from a number, the number is not changed.

Examples $6 - 0 = 6$ $1{,}999 - 0 = 1{,}999$

Times 0: If a number is multiplied by 0, the answer is 0.

Examples $6 \times 0 = 0$ $0 \times 46 = 0$ $1{,}999 \times 0 = 0$

Times 1: If a number is multiplied by 1, the number is not changed.

Examples $1 \times 6 = 6$ $46 \times 1 = 46$ $1 \times 812 = 812$

Times 5: To multiply by 5, think "nickels."

Example $7 \times 5 = ?$ 7 nickels is 35¢. So $7 \times 5 = 35$.

Times 10: To multiply by 10, think "dimes."

Example $7 \times 10 = ?$ 7 dimes is 70¢. So $7 \times 10 = 70$.

Turn-around shortcut for addition: Numbers have the same sum when they are added in reverse order.

Examples $4 + 9 = 9 + 4$ $135 + 60 = 60 + 135$

Turn-around shortcut for multiplication: Numbers have the same product when they are multiplied in reverse order.

Examples $6 \times 9 = 9 \times 6$ $105 \times 41 = 41 \times 105$

Partial-Sums Addition Method

There are different methods you can use to add. One of these is called the **partial-sums method.** It is described below. You may have a favorite addition method of your own. Even if you do, make sure that you can also use the partial-sums method.

Here is the partial-sums method for adding 2-digit or 3-digit numbers:

1. Add the 100s.

2. Add the 10s.

3. Add the 1s.

4. Add the sums you just found (the partial sums).

Did You Know?

Another word for *method* is *algorithm*. A method is a clear set of rules used to solve a problem. So is an algorithm.

Example Add 248 + 187 using the partial-sums method.

		100s	10s	1s
		2	**4**	**8**
		+ **1**	**8**	**7**
Add the 100s.	200 + 100 →	3	0	0
Add the 10s.	40 + 80 →	1	2	0
Add the 1s.	8 + 7 →		1	5
Add the partial sums.	300 + 120 + 15 →	**4**	**3**	**5**

248 + 187 = 435

Numbers with 4 or more digits are added in the same way.

Check Your Understanding

Use the partial-sums method to add.

1. 34 + 62 **2.** 34 + 88 **3.** 123 + 456 **4.** 408 + 393

Check your answers on page 337.

You can use base-10 blocks to show how the partial-sums addition method works.

Example Use base-10 blocks to add 248 + 187.

Each base-10 cube is worth 1.

Each base-10 long is worth 10.

And each base-10 flat is worth 100.

Add the blocks in each column. Then find the total.

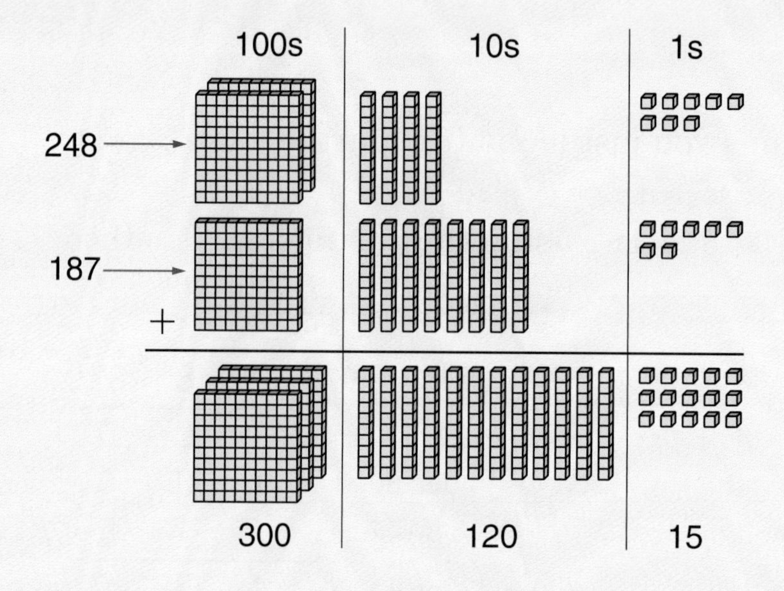

Find the total. 300 + 120 + 15 = 435

248 + 187 = 435

Column-Addition Method

Many people prefer the **column method** for adding.

Here is the column method for adding 2-digit or 3-digit numbers:

1. Draw lines to separate the 1s, 10s, and 100s places.
2. Add the numbers in each column. Write each sum in its column.
3. If there are 2 digits in the 1s place, trade 10 ones for 1 ten.
4. If there are 2 digits in the 10s place, trade 10 tens for 1 hundred.

Example Add 248 + 187 using the column-addition method.

	100s	10s	1s
	2	**4**	**8**
	+ **1**	**8**	**7**
Add the numbers in each column.	3	12	15
Two digits in the ones place. Trade 15 ones for 1 ten and 5 ones. Move the 1 ten to the tens column.	3	13	5
Two digits in the tens place. Trade 13 tens for 1 hundred and 3 tens. Move the 1 hundred to the hundreds column.	**4**	**3**	**5**

248 + 187 = 435

Trade-First Subtraction Method

One method you can use to subtract is called the **trade-first method.** Here is the trade-first method for subtracting 2-digit or 3-digit numbers:

1. Look at the digits in the 1s place. If subtracting these digits gives a negative number, trade 1 ten for 10 ones.
2. Look at the digits in the 10s place. If subtracting these digits gives a negative number, trade 1 hundred for 10 tens.
3. Subtract in each column.

Example Subtract 164 from 352 using the trade-first method.

	100s	10s	1s
	3	5	2
−	1	6	4

Look at the 1s place. You cannot remove 4 ones from 2 ones.

	100s	10s	1s
		4	12
	3	~~5~~	~~2~~
−	1	6	4

So trade 1 ten for 10 ones. Now look at the 10s place. You cannot remove 6 tens from 4 tens.

	100s	10s	1s
		14	
	2	~~4~~	12
	~~3~~	~~5~~	~~2~~
−	1	6	4
	1	**8**	**8**

So trade 1 hundred for 10 tens. Now subtract in each column.

352 − 164 = 188

Check Your Understanding

Use the trade-first method to subtract.

1. 67 − 29 **2.** 132 − 115 **3.** 248 − 72 **4.** 306 − 155

Check your answers on page 337.

Base-10 blocks are useful for solving problems. If these blocks are not handy, you can draw pictures instead.

Shorthand pictures are used for this example of the trade-first method.

Base-10 Blocks and Their Shorthand Pictures

cube long flat

Example $324 - 167 = ?$

Use pictures of base-10 blocks to model the larger number, 324.
Write the number to be subtracted, 167, beneath the block pictures.

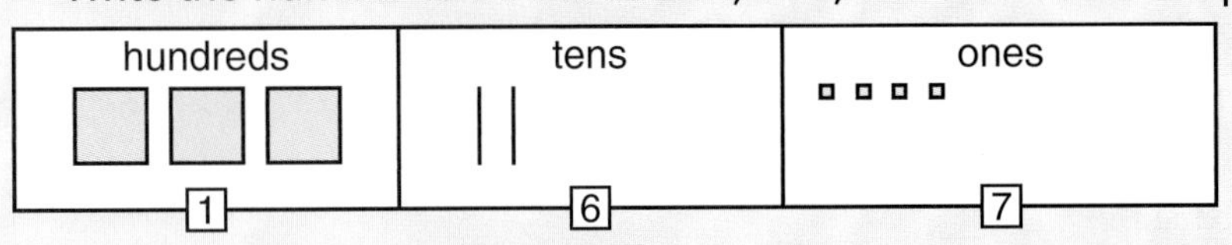

100s	10s	1s
3	2	4
– 1	6	7

Think: Can I remove 7 cubes from 4 cubes? No. Trade 1 long for 10 cubes.

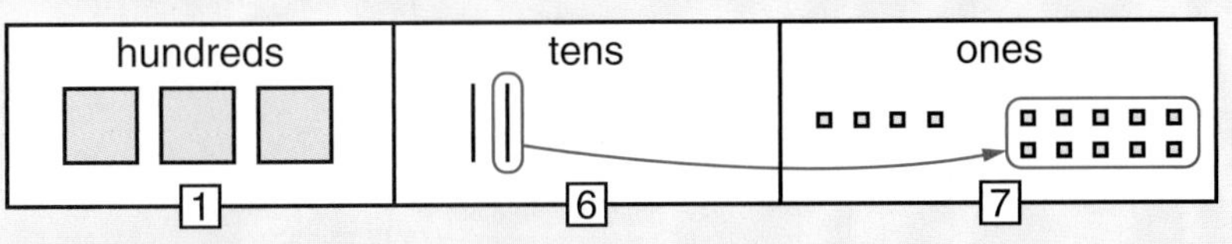

Think: Can I remove 6 longs from 1 long? No. Trade 1 flat for 10 longs.

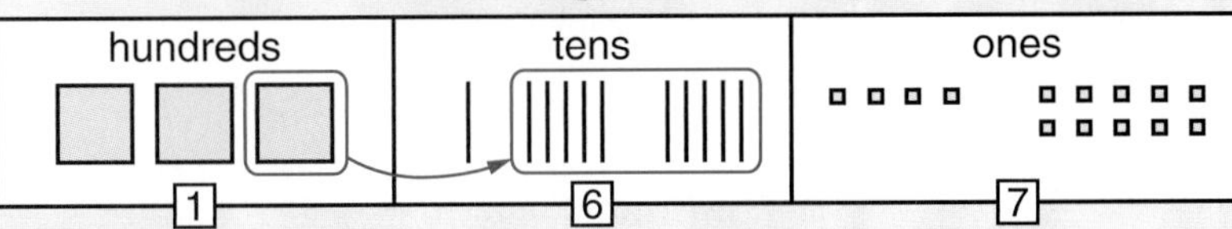

After all of the trading, the blocks look like this:

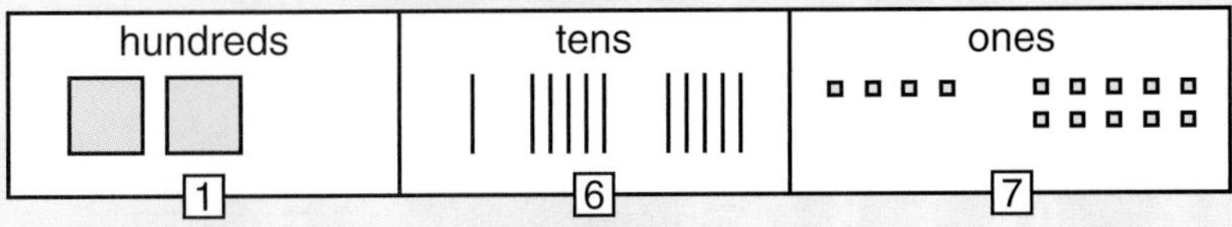

Now subtract in each column.

hundreds tens ones

100s	10s	1s
3	2	4
– 1	6	7
1	5	7

The difference is 157.

Left-to-Right Subtraction Method

You can subtract two numbers by subtracting one column at a time. Start with the left column and end with the right column. That is why the method is called the **left-to-right method.**

Here is the left-to-right method for subtracting 2-digit or 3-digit numbers:

1. Subtract the 100s.

2. Next subtract the 10s.

3. Then subtract the 1s.

Examples Use the left-to-right method to solve these problems.

$$\begin{array}{r} 60 \\ -\ 27 \\ \hline \end{array}$$

There are no hundreds.
So subtract the 10s first.

		10s	1s
		6	**0**
Subtract the 10s.	−	**2**	0
		4	0
Subtract the 1s.	−		**7**
		3	**3**

60 − 27 = 33

$$\begin{array}{r} 932 \\ -\ 356 \\ \hline \end{array}$$

		100s	10s	1s
		9	**3**	**2**
Subtract the 100s.	−	**3**	0	0
		6	3	2
Subtract the 10s.	−		**5**	0
		5	8	2
Subtract the 1s.	−			**6**
		5	**7**	**6**

932 − 356 = 576

Counting-Up Subtraction Method

You can subtract two numbers by counting up from the smaller number to the larger number. Subtracting this way is called the **counting-up method.**

1. Write the smaller number. Count up to the nearest multiple of 10.
2. Keep counting up by 10s and 100s.
3. Then count up to the larger number.

Example Subtract 38 from 325 by counting up. Write the smaller number, 38, and count up to 325. Circle each number that you count up.

```
     3 8
+     (2)     Count up to the nearest 10.
--------
     4 0
+   (6 0)     Count up to the nearest 100.
--------
   1 0 0
+ (2 0 0)     Count up to the largest
--------      possible hundred.
   3 0 0
+   (2 5)     Count up to the larger number.
--------
   3 2 5
```

Then, add the numbers you circled: $2 + 60 + 200 + 25 = 287$
You counted up by 287.

$325 - 38 = 287$

Check Your Understanding

Use the left-to-right or counting-up method to subtract.

1. $90 - 33$ **2.** $242 - 70$ **3.** $835 - 451$ **4.** $520 - 148$

Check your answers on page 337.

Arrays

An **array** is a group of objects arranged in **rows** and **columns.**

- The outline around the rows and columns is a rectangle.
- Each row is filled and has the same number of objects.
- Each column is filled and has the same number of objects.

Example The push buttons on a telephone are an array.

The outline around the buttons is a rectangle.
There are 3 buttons in each row.
There are 4 buttons in each column.

This array is called a 4-by-3 array.

Example The eggs in this carton are an array.

Here is one way to describe the array:

- It has 2 rows, with 6 eggs in each row.
- It is called a 2-by-6 array.

Here is another way to describe the array:

- It has 6 columns, with 2 eggs in each column.
- It is called a 6-by-2 array.

Example There are 4 different ways to make an array that has 10 objects.

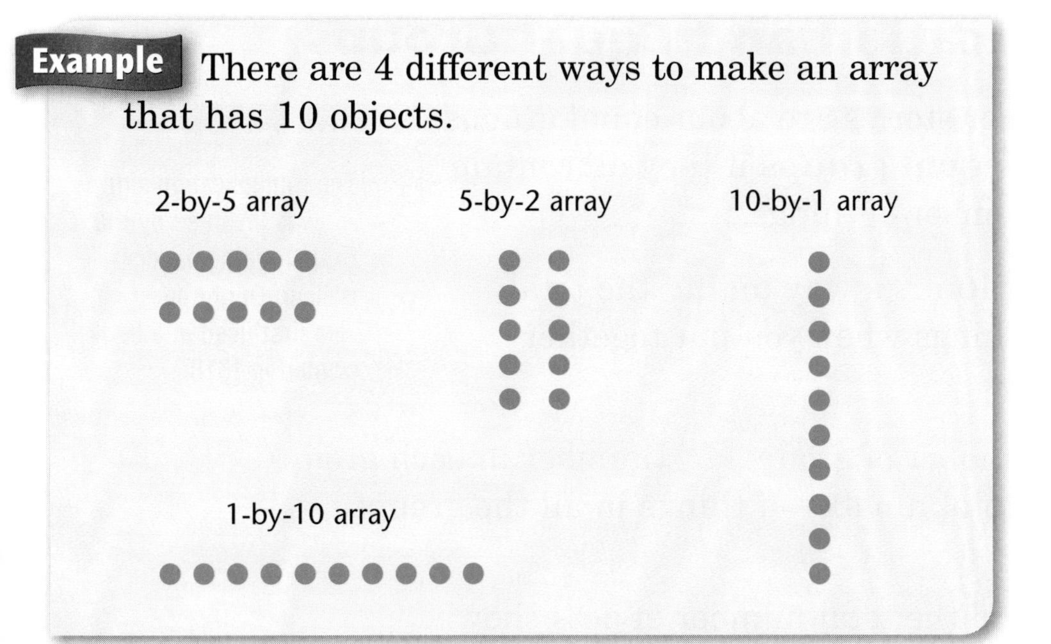

Arrays are useful for showing **equal groups** of objects. The groups are equal groups if they all contain the same number of objects.

Example Louise buys 4 packages of juice boxes. Each package has 6 juice boxes in it and is called a "six-pack." Show the 4 packages as an array.

There are 4 groups (4 packages of juice boxes).
Each group has the same number of boxes (6) in it.

Draw a 4-by-6 array to show the 4 groups.

Each row stands for 1 package (or six-pack).

Multiplication and Equal Groups

Many number stories are about equal groups. Groups are **equal groups** if they all contain the same number of things.

Multiplication is a way to find the total number of things when you put together equal groups.

Did You Know?

The multiplication sign, ×, was invented by the British mathematician William Oughtred. It was first used in a book printed in 1618.

Multiply (number of groups) by (number in each group) to find the total number of things in all the groups.

Example Find the total number of dots shown.

One way to find the total number of dots is to count them. There are 12 dots.

Another way to find the total number of dots is to think of equal groups of dots. There are 3 groups and 4 dots in each group. So this is an equal-groups problem.

To find the total number of dots,
multiply (number of groups) × (number in each group): $3 \times 4 = 12$.
3 groups of 4 is the same as 12.

There are 12 dots in all.

We say that $3 \times 4 = 12$ is a number model for the problem.

The × is a **multiplication sign.** Read 3×4 as "3 **times** 4" or as "3 **multiplied by** 4."

Arrays are very useful for showing equal groups of objects. Drawing arrays can help you solve problems involving equal groups.

Example There are 6 cartons with 4 bottles per carton. How many bottles are there in all?

The word **per** means "in each." So, there are 4 bottles in each carton.

The 6 cartons are the 6 groups. Each carton has the same number of bottles (4). So this is an equal-groups problem.

You can draw an array to show the 6 equal groups. Each row stands for 1 carton with 4 bottles in it.

Multiply (number of rows) by (number of columns) to find the total number of objects in the array: $6 \times 4 = 24$.
6 groups of 4 is the same as 6×4, or 24.

We write 6×4 to show 6 groups of 4.
There are 24 bottles in all.

A number model for the problem is $6 \times 4 = 24$.

Check Your Understanding

1. There are 4 tables and 8 chairs per table. How many chairs are there in all? Draw an array to help you solve the problem.
2. Write a multiplication number model for each array.

a.

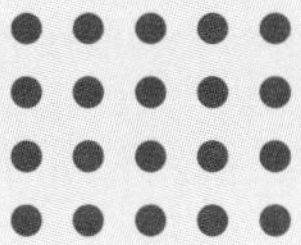

b.

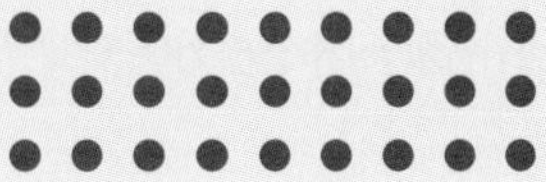

Check your answers on page 338.

Partial-Products Multiplication Method

One way to multiply numbers is called the **partial-products method.** Write 1s, 10s, and 100s above the columns, as shown below.

Example Multiply 5×26.

Think of 26 as 2 tens and 6 ones.

Multiply each part of 26 by 5.

		100s	10s	1s
			2	**6**
		$\times$		**5**
5 ones $\times$ 2 tens:	$5 \times 20 \rightarrow$	1	0	0
5 ones $\times$ 6 ones:	$5 \times 6 \rightarrow$		3	0
Add these two parts:	$100 + 30 \rightarrow$	**1**	**3**	**0**

$5 \times 26 = 130$

The next example uses an array diagram to solve the same problem.

Example Use an array diagram to show 5×26.

Draw an array to show 5 rows with 26 in each row.
Divide each row to show 2 tens (blue dots) and 6 ones (red dots).

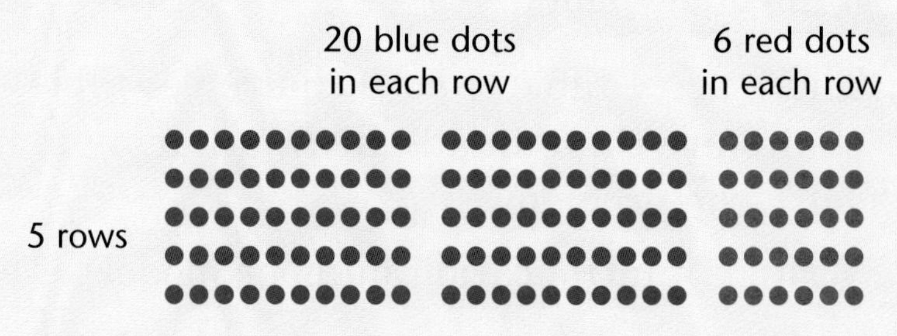

Multiply each part of 26 by 5: $5 \times 20 = 100$ $5 \times 6 = 30$

Add these two parts: $5 \times 26 = 100 + 30$, which is 130.

$5 \times 26 = 130$

Example Multiply 34×26.

		100s	10s	1s
Think of 26 as 2 tens and 6 ones.			**2**	**6**
Think of 34 as 3 tens and 4 ones.	$\times$		**3**	**4**
Multiply each part of 26 by each part of 34.				
3 tens $\times$ 2 tens:	$30 \times 20 \rightarrow$	6	0	0
3 tens $\times$ 6 ones:	$30 \times 6 \rightarrow$	1	8	0
4 ones $\times$ 2 tens:	$4 \times 20 \rightarrow$		8	0
4 ones $\times$ 6 ones:	$4 \times 6 \rightarrow$		2	4
Add these four parts:	$600 + 180 + 80 + 24 \rightarrow$	**8**	**8**	**4**

$34 \times 26 = 884$

The problem 34×26 has been split up into four easy problems. The answer to each easy problem is called a **partial product.** Adding these partial products gives the answer to 34×26.

Check Your Understanding

Use the partial-products method to multiply.

1. 6×37

2. $\begin{array}{r} 58 \\ \times \quad 7 \\ \hline \end{array}$

3. $\begin{array}{r} 1{,}574 \\ \times \qquad 7 \\ \hline \end{array}$

4. $\begin{array}{r} 69 \\ \times 34 \\ \hline \end{array}$

Check your answers on page 338.

Lattice Multiplication Method

The **lattice method** for multiplying numbers has been used for hundreds of years. It is very easy to use if you know the basic multiplication facts. Study the examples below.

Example Use the lattice method to multiply 3×45.

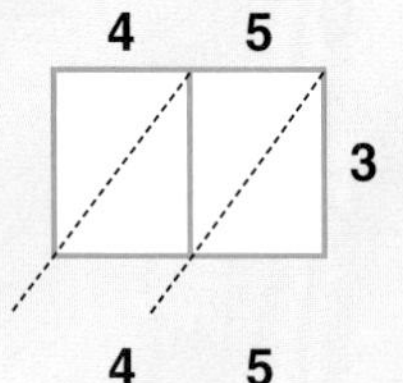

The box with squares and diagonals is called a **lattice.**
Write 45 above the lattice.
Write 3 on the right side of the lattice.

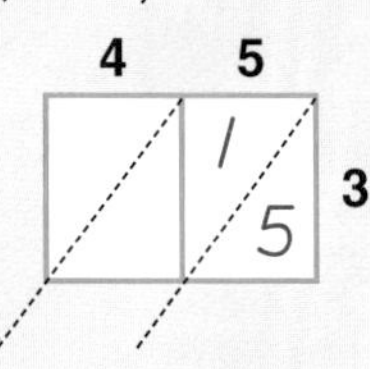

Multiply 3×5.
Write the answer as shown.

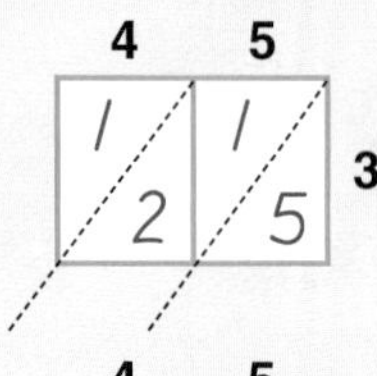

Multiply 3×4.
Write the answer as shown.

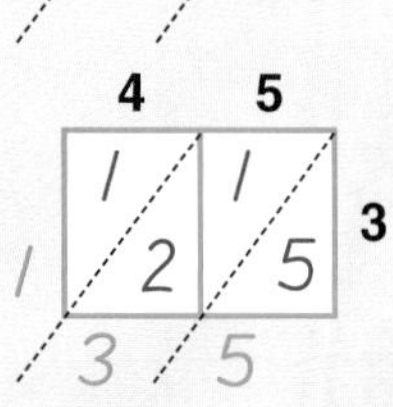

Add the numbers along each diagonal, starting at the right.

Read the answer. $3 \times 45 = 135$

Example Use the lattice method. Multiply 4×713.

$4 \times 713 = 2{,}852$

The numbers along a diagonal may add up to a 2-digit number. When this happens . . .

- Write the 1s digit.
- Add the 10s digit to the sum along the diagonal above.

Example Use the lattice method. Multiply 7×89.

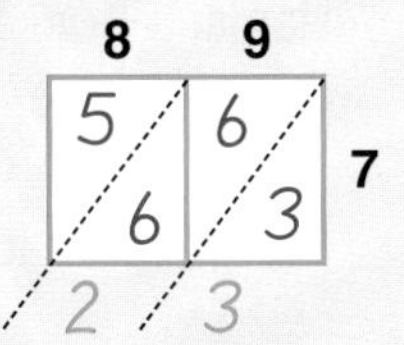

The sum along one diagonal is $6 + 6 = 12$. Write the 1s digit of 12.

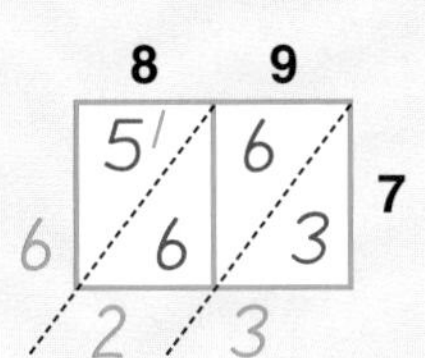

Add the 10s digit of 12 to the sum along the diagonal above. This sum is 6.

Read the answer. $7 \times 89 = 623$

Example Use the lattice method. Multiply 34×26.

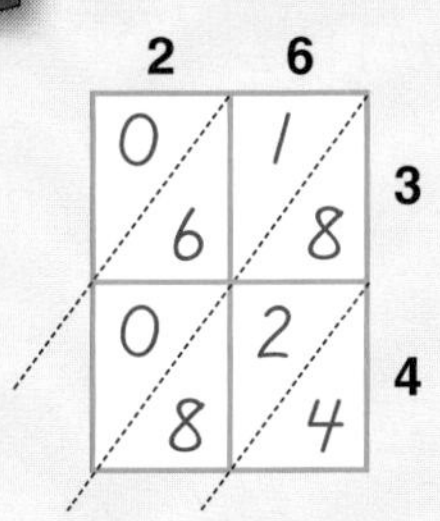

Write 26 above the lattice.
Write 34 on the right side of the lattice.

Multiply 3×6. Then multiply 3×2.
Multiply 4×6. Then multiply 4×2.

Write the answers as shown in the lattice.

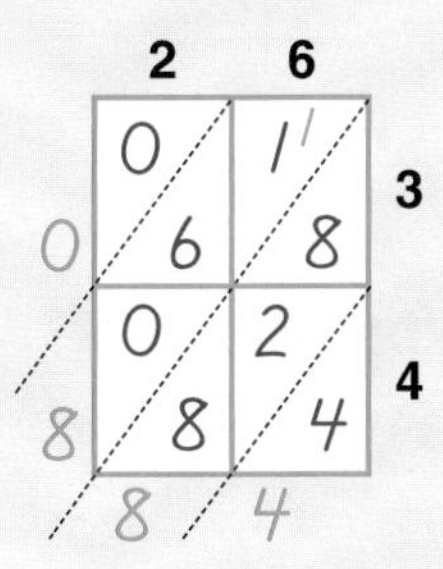

Add the numbers along each diagonal starting at the right.

For the sum 18, write 8. Then add 1 to the sum along the diagonal above.

Read the answer. $34 \times 26 = 884$

The search for ways to record computation started in India, perhaps about the eleventh century. The lattice method of multiplication was probably passed on from the Hindus to the Arabians. The Arabians then passed it on to the Europeans. Fifteenth-century writers in western Europe included it in their printed books.

The first printed arithmetic book appeared in Italy in 1478. Luca Pacioli listed eight different ways to do multiplication in this book. He called one of the ways "lattice multiplication." The name suggests the gratings that were placed in windows to keep people from looking through them.

Did You Know?

In the late 1500s, John Napier invented a set of numbered rods that could be used to multiply numbers. The rods were strips of wood or bone, called Napier's Bones. Using these rods to multiply is almost the same as using the lattice method.

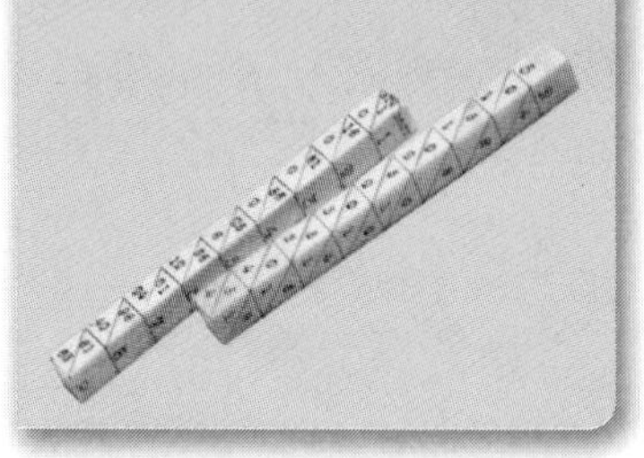

Check Your Understanding

1.

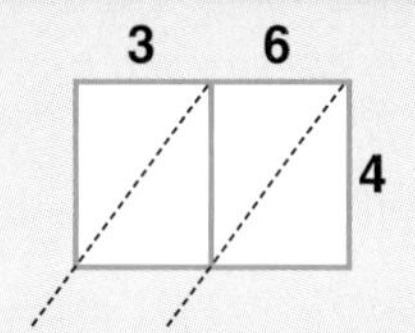

$4 \times 36 =$ ___

2.

5 1 7
3

$3 \times 517 =$ ___

3.

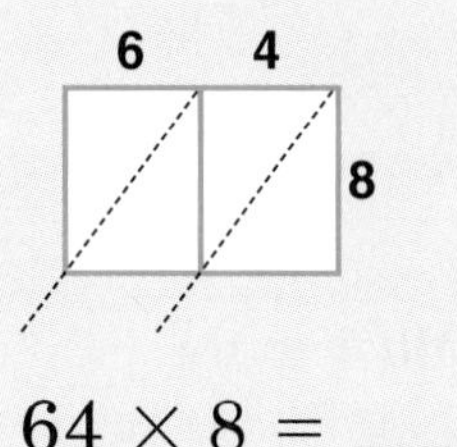

$64 \times 8 =$ ___

4.

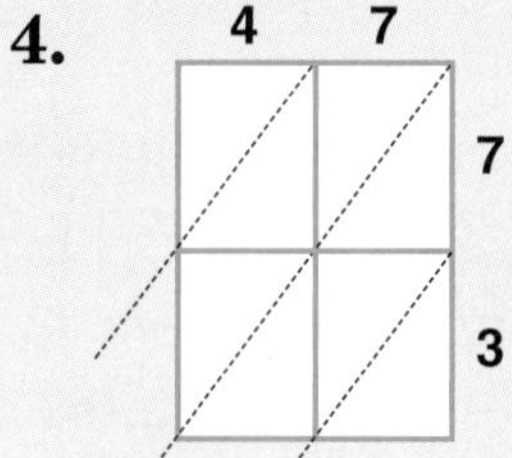

$47 \times 73 =$ ___

Check your answers on page 338.

Division and Equal Sharing

An **equal-sharing story** involves dividing a group of things into equal parts. Equal parts are also called equal shares.

Equal-sharing stories are also called **division stories.**

Sometimes it is not possible to divide a group into equal shares. The number left over is called the **remainder.**

The **division sign** (÷) is used to show division. It is used in writing number models for equal-sharing stories.

Example Four boys share 24 marbles equally. What is each boy's share?

You can divide the 24 marbles into 4 equal shares.

Each share has 6 marbles. The remainder is 0 because no marbles are left over.

A number model for the problem is $24 \div 4 = 6$. Read this as "24 divided by 4 equals 6."

Example Four boys share 26 marbles equally. Each share has 6 marbles. The remainder is 2 since 2 marbles are left over. A number model for the problem is $26 \div 4 = 6$ (remainder 2).

Division and Equal Grouping

Some division stories involve making equal groups of things. The number of things in any one group is known. The problem is to find the number of groups that can be made from the things you have.

Example 23 children want to play ball. How many teams can you make with 5 children per team?

The word **per** means "for each."
So there are 5 children for each team.
Each team of 5 children is one group.

You can use counters or tallies to find how many groups of 5 can be made with 23 children.

𝍸
𝍸
𝍸
𝍸
///

The tallies above show that 23 children can be divided into 4 groups (teams), with 3 children left over. The remainder is 3.

A number model for this number story is
$23 \div 5 = 4$ (remainder 3).

Summary for division and equal sharing:

Divide (total) by (number of shares) to find the number in any 1 share.

Summary for division and equal grouping:

Divide (total) by (number in 1 group) to find the number of groups that can be made from the total.

Data and Chance

Tally Charts

There are different ways you can collect information about something.

- Count
- Measure
- Ask questions
- Look at something, and describe what you see

The information you collect is called **data.** You can use a **tally chart** to record and organize data.

Example Mr. Davis asked each student to name his or her favorite drink. He recorded the students' choices in the tally chart below.

Favorite Drinks

Drink	Tallies
Milk	~~////~~
Chocolate milk	///
Soft drink	~~////~~ ~~////~~ /
Apple juice	///
Tomato juice	/
Water	//

Milk (5 votes) is more popular than chocolate milk (3 votes).

Soft drink is the most popular choice (11 votes).

Tomato juice is the least popular choice (1 vote).

There are 25 tally marks in the chart.
That means that 25 different students voted for a favorite drink.

Check Your Understanding

1. How many children voted for apple juice?
2. Which drinks are less popular than apple juice?

Check your answers on page 338.

Tally Charts and Line Plots

A **line plot** is another way to organize data and make it easier to understand. The example below shows that line plots are very much like tally charts.

Example Mr. Ramirez gave his class a 5-word spelling test. Here are the children's scores (the number correct):

Ann 3	Joe 4	Britney 2	Lilly 4
Stan 4	Hanna 3	Jim 5	Carlos 4
Tanesha 1	Ramon 2	Ari 1	Ted 3
Aaron 3	Dina 2	Mark 4	Tina 4

The test scores can be organized in a tally chart or in a line plot. Both ways use marks to show how many children got each score.

The tally chart uses tally marks.

Test Scores

Number Correct	Tallies
0	
1	//
2	///
3	////
4	~~////~~ /
5	/

The line plot uses Xs.

Test Scores

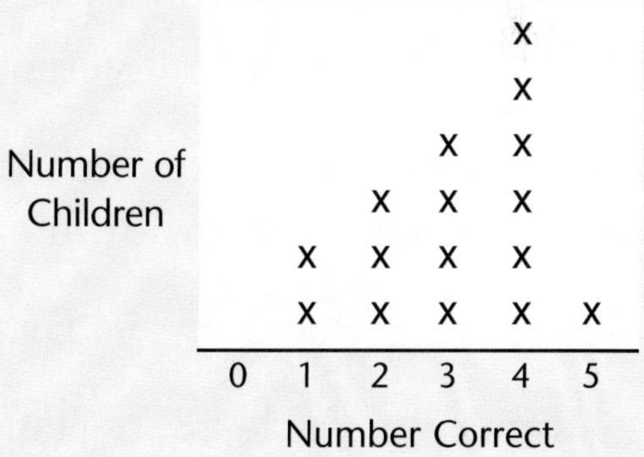

Number of Children						
					X	
					X	
				X	X	
			X	X	X	
		X	X	X	X	
		X	X	X	X	X
Number Correct	0	1	2	3	4	5

The tally chart and line plot both show that 6 children got 4 words correct.

Think of turning the line plot one quarter turn ↘. The tally chart and line plot will then look about the same.

Check Your Understanding

1. Ms. Clark's class is having a picnic. Class members decided on the day for the picnic by voting.

The tally chart shows how the class voted.

Votes on Day for a Picnic

Day	Tallies
Monday	~~||||~~ |
Tuesday	
Wednesday	||||
Thursday	~~||||~~ ||
Friday	~~||||~~ |||

a. Which day got the most votes?

b. Which day got the least votes?

c. How many children voted for Wednesday?

d. How many children voted?

2. Mr. Ramirez gave some children another 5-word spelling test. Here are the children's scores (the number correct):

Ann 3	Joe 5	Britney 4	Lilly 5	Stan 1
Jim 4	Carlos 3	Tanesha 3	Ramon 3	Ari 5

Make a line plot to organize the test scores.

Check your answers on page 338.

Describing a Set of Data: The Minimum, Maximum, and Range

If you were asked to describe a certain car, you would probably talk about some of its more important features. You might say, "It is a 2007 midsize car. It is red and has 4 doors. It has 2 front and 2 side air bags. It has been driven about 25,000 miles."

If you were asked to describe the numbers in a data set, you might mention these features:

- The **minimum** is the smallest number.
- The **maximum** is the largest number.
- The **range** is the difference between the largest and the smallest numbers.

Did You Know?

The highest (maximum) temperature ever recorded on Earth was 136°F, in Libya. The lowest (minimum) temperature ever recorded was −129°F, in Antarctica.

Example Andrew kept a record of the number of pages he read each day.

Mon	Tue	Wed	Thu	Fri
27	15	12	20	18
↑ maximum (largest) number: 27 pages		↑ minimum (smallest) number: 12 pages		

To find the **range,** subtract the smallest number from the largest number.

The range is 27 − 12, or 15 pages.

Describing a Set of Data: The Median

The numbers in a set of data are often arranged in order. They can be listed from smallest to largest or from largest to smallest. The **median** is the number in the middle of the list. The median is also known as the **middle number** or the **middle value.**

Did You Know?

On some roads and highways, there is a dividing area between the opposite lanes of traffic. This middle area is called a *median,* or a *median strip.*

Example What is the median height of these five children?

Ricky	Marla	Suki	Alan	Dan
48 inches	52 inches	51 inches	45 inches	50 inches

List the numbers in order. The middle number is 50.

45 48 **50** 51 52

So the median height is 50 inches.

Example What is the median weight of these six children?

Child	Weight
Mary	63 pounds
Ravi	58 pounds
Carl	53 pounds
Yoko	66 pounds
Sue	44 pounds
Eddie	56 pounds

List the numbers in order. There are two middle numbers. The median is the number halfway between these middle numbers.

44 53 **56** **58** 63 66

So the median weight is 57 pounds.

Describing a Set of Data: The Mode

When you study a set of data, you may notice that one number or answer occurs most often. The **mode** is the number or answer that occurs most often.

Example Sue kept a record of the number of pages she read each day.

Mon	Tue	Wed	Thu	Fri
13	20	25	16	16

The number 16 is listed twice (on Thursday and Friday). The other numbers are listed only one time each.

The mode is 16 pages.

If you make a tally chart or a line plot, the mode is easy to find.

Examples

Our Class's Favorite Drinks

Drink	Tallies
Milk	~~////~~
Chocolate milk	///
Soft drink	~~////~~ ~~////~~ /
Apple juice	///
Tomato juice	/
Water	//

The answer given most often was "soft drink." So the mode is soft drink.

Scores on a 5-Word Test

Number of Children

				X	
				X	
			X	X	
		X	X	X	
	X	X	X	X	
	X	X	X	X	X
0	1	2	3	4	5

Number Correct

The score given most often was 4. So the mode is 4.

Check Your Understanding

1. Brad kept a record of the number of minutes he did homework each day.

Mon	Tue	Wed	Thu	Fri
45	23	42	40	31

Find the minimum, maximum, and range for this set of data.

2. Here are the points that six basketball players scored:

Player	Points
1	7
2	3
3	7
4	5
5	0
6	10

   a. Find the minimum, maximum, and range for this set of data.
   b. Find the median number of points scored.
   c. Find the mode for this set of data.

3. Stephanie kept a record of her math quiz scores. Here is her list of quiz scores:

10 7 6 9 6 10 9 8 5 9 6 6 8 8 7

What is the mode for these quiz scores?

Check your answers on pages 338 and 339.

The Mean (Average)

Here are three stacks of pancakes:

This is not fair.

One person gets 6 pancakes and another gets 2.

There are 12 pancakes in all. We can move some pancakes to make the stacks equal. Then each stack will have 4 pancakes.

This is fair.

Each person gets 4 pancakes.

We say that 4 is the **mean** number of pancakes in each stack.

Here is how to find the mean:

Step 1: Find the total amount in all the groups.

Step 2: Find the amount that would be in each group if the groups were equal.

The mean is sometimes called the **average.**

Did You Know?

Average was first used as an English word around 1500. One of its meanings was "an equal share." When an expense was shared equally by a group of people, each person paid the average.

Example Jacob earned \$4 and Emma earned \$2. What is the mean amount they earned?

Jacob Emma

Step 1: Find the total: \$4 + \$2 = \$6.

Jacob Emma

Step 2: Divide the total (\$6) by the number of groups (2) to make equal groups: \$6 ÷ 2 = \$3.

The mean (average) amount Jacob and Emma earned is \$3.

If Jacob and Emma share their earnings equally, each will get \$3.

Example Five children take a hike. Each child carries a backpack. Find the mean weight of their backpacks.

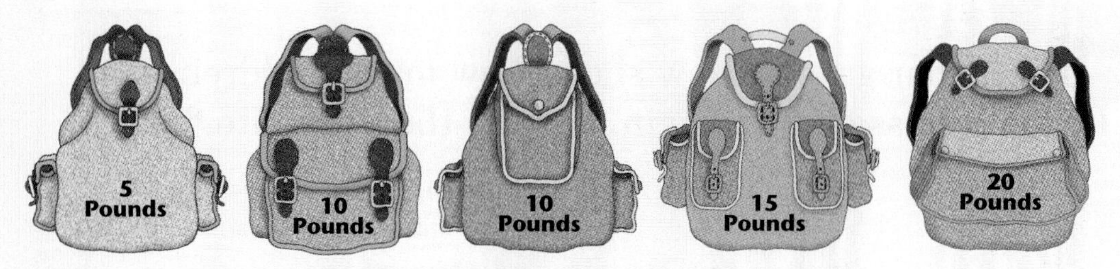

Step 1: Add to find the total weight.
$5 + 10 + 10 + 15 + 20 = 60$ pounds

Step 2: Divide the total weight (60 pounds) by the number of backpacks (5).
60 pounds $\div$ 5 = 12 pounds

The mean (average) weight of the backpacks is 12 pounds.

If the children rearrange the items in the backpacks so that all 5 backpacks have the same weight, each backpack will weigh 12 pounds.

Check Your Understanding

1. Glen has 20 model cars. Chad has 13 model cars. Tom has 15 model cars. What is the mean (average) number of cars that the boys have?
2. Olivia kept a record of her math quiz scores. What is Olivia's mean score?

Quiz #1	Quiz #2	Quiz #3	Quiz #4
10	5	10	7

Check your answers on page 339.

Bar Graphs

A **bar graph** is a drawing that uses bars to show numbers.

Example The bar graph below shows how many children in a Grade 3 class chose certain foods as their favorite foods.

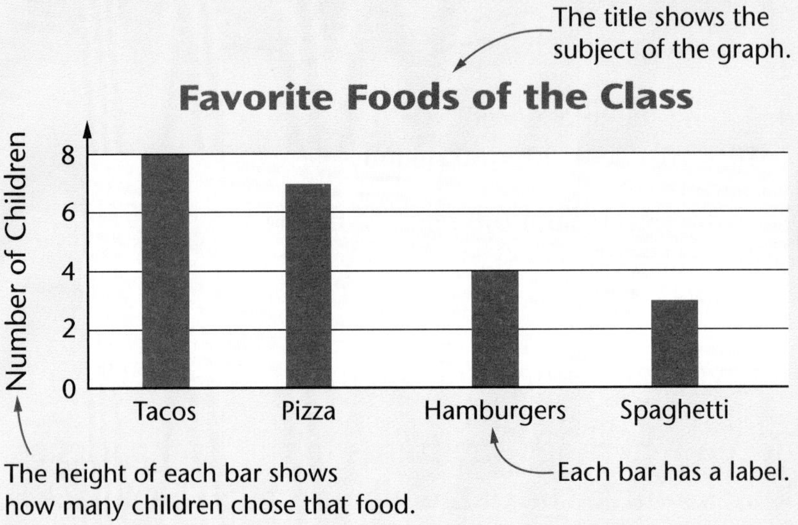

The title shows the subject of the graph.

Favorite Foods of the Class

Number of Children

8
6
4
2
0

Tacos Pizza Hamburgers Spaghetti

The height of each bar shows how many children chose that food.

Each bar has a label.

You can answer questions:

How many children chose pizza?

The bar for pizza ends halfway between the line for 6 and the line for 8. So, pizza was the favorite food of 7 children.

You can compare choices:

Eight children chose tacos as their favorite food. Only 3 children chose spaghetti. Tacos are more popular than spaghetti.

Did You Know?

The first bar graph was drawn in 1786 by William Playfair. Bar graphs today are one of the most popular types of graph for showing and comparing data.

When data are collected, sometimes they are put in a tally chart before a bar graph is made.

Example The children in a Grade 3 class counted how many pull-ups each of them could do. Their results are shown in the tally chart.

Number of Pull-Ups	Number of Children
0	~~////~~ /
1	~~////~~
2	////
3	//
4	
5	///
6	/

The bar graph below shows the same information as the tally chart, but in a different way.

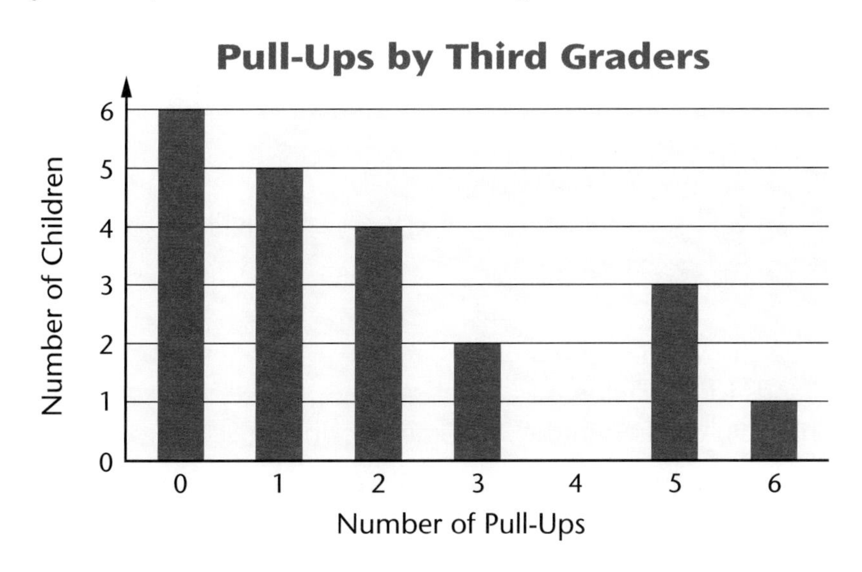

Pictographs

A **pictograph** uses picture symbols to show numbers.

Example The pictograph below shows how many children chose certain foods as their favorite foods.

Favorite Foods of the Class

Tacos

Pizza

Hamburgers

Spaghetti

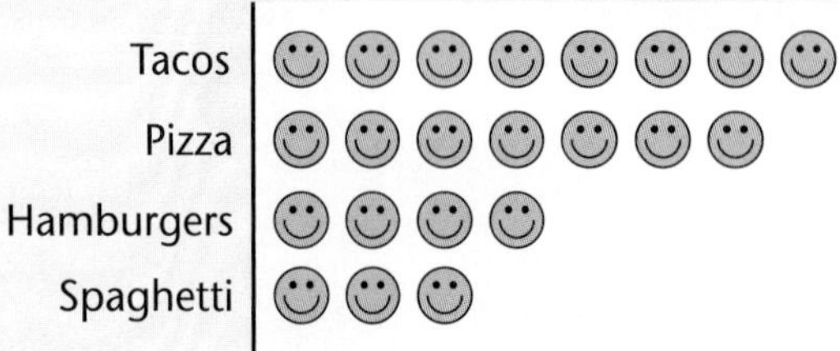

KEY: ☺ = 1 child

The KEY tells you what each picture symbol is worth.

The line for tacos shows 8 face symbols.
Each face symbol stands for 1 child.
So, 8 children chose tacos as their favorite food.

If each ☺ symbol is replaced by 1 tally mark, the pictograph will become a tally chart.

When you use a pictograph, always check the KEY first.

Example This pictograph shows how many children in Lincoln School are in each grade.

Number of Children in Each Grade

3rd Grade

4th Grade

5th Grade

KEY: ☺ = 10 children

The line for 3rd Grade shows 9 face symbols.
Each face symbol stands for 10 children.
So, there are $9 \times 10 = 90$ children in the 3rd Grade of Lincoln School.

If each ☺ symbol is replaced by 10 tally marks ~~||||~~ ~~||||~~, the pictograph will become a tally chart.

In some pictographs, you may see only part of a picture symbol. Use the KEY to decide how much this part of the symbol is worth.

Example This pictograph shows how many children in each grade of Lincoln School ride a bicycle to school.

Number of Children Who Ride a Bicycle to School

3rd Grade

4th Grade

5th Grade

KEY: = 2 children

stands for 2 children. So, stands for 1 child.

$2 + 1 = 3$ children in the 3rd Grade ride a bicycle to school.
$2 + 2 + 2 = 6$ children in the 4th Grade ride a bicycle to school.
$2 + 2 + 2 + 2 + 1 = 9$ children in the 5th Grade ride a bicycle to school.

Example Kellog School held a weekend car wash. How many cars were washed in all?

Number of Cars Washed

Friday

Saturday

Sunday

KEY: = 6 cars

stands for 6 cars. So, stands for 3 cars.

$2 \times 6 = 12$ cars were washed on Friday.
$(4 \times 6) + 3 = 24 + 3 = 27$ cars were washed on Saturday.
$(3 \times 6) + 3 = 18 + 3 = 21$ cars were washed on Sunday.
$12 + 27 + 21 = 60$ cars were washed in all.

Line Graphs

A line graph is often used to show how something has changed over a period of time.

Example At 2:00 P.M. each day for a week, Ashley checked her outdoor thermometer and read the temperature. She recorded her information on a line graph.

Outdoor Temperature at 2:00 P.M.

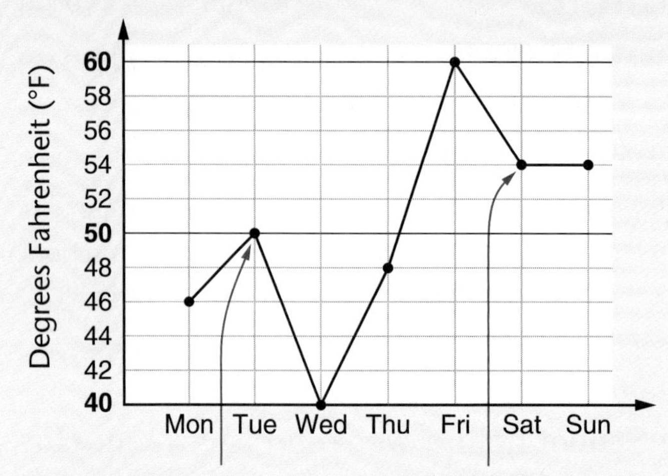

The temperature was 50°F at 2:00 P.M. Tuesday.

The temperature was 54°F at 2:00 P.M. Saturday.

By carefully reading the graph, you can get a lot of information.

- The graph shows that it was warmer on weekend days (Friday, Saturday, and Sunday) than on other days.
- It was 20 degrees warmer on Friday than it was on Wednesday.

Check Your Understanding

1. Use the line graph on page 90 to answer the questions.
   a. What was the highest temperature?
   b. What was the lowest temperature?
   c. What days had the same temperature?
   d. About what temperature would describe the temperature for the week? Explain your answer.
   e. On how many days did Ashley record the temperature?

2. Nine players played in a baseball game. The table shows how many hits each player made. Make a bar graph to show this information. Copy the following grid on a separate sheet of paper. Then draw the bars.

Players	Hits
Carl	1
Mary	3
Laci	1
Jamil	0
Lee	2
Ed	4
Ali	1
Tanya	0
Nancy	2

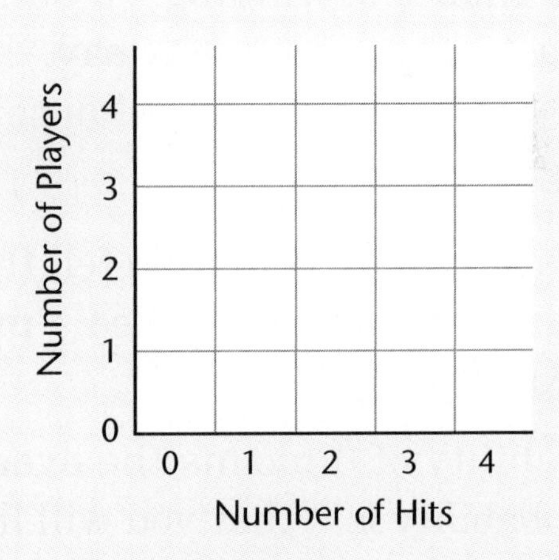

Check your answers on page 339.

Chance and Probability

Things that happen are called **events.** There are many events that you can be sure about.

- You are **certain** that the sun will rise tomorrow.
- It is **impossible** for you to grow to be 12 feet tall.

There are also many events that you *cannot* be sure about.

- You cannot be sure whether it will be sunny or cloudy next Friday.
- You cannot be sure that you will get a letter tomorrow.

You often talk about the **chance** that something will happen. If Pam is a fast runner, you may say, "Pam has a good chance of winning." If Jan is also a fast runner, you may say, "Pam and Jan have the same chance of winning."

Sometimes a number is used to tell the chance of something happening. This number is called a **probability.** It is a number from 0 to 1.

- A probability of 0 means the event is *impossible.* The probability is 0 that you will live to the age of 150.
- A probability of 1 means that the event is *certain* to happen. The probability is 1 that the sun will rise tomorrow.

A probability of $\frac{1}{2}$ means that the event will happen about half of the time.

Example What do you expect will happen if you spin this spinner many times?

Half of the spinner is red, and half is green. So your chance of landing on red is the same as your chance of landing on green.

The red and green sections are **equally likely.** They have the same chance of being landed on.

If you spin the spinner many times, it will land on red about 1 out of 2 spins, or about half of the time.

The probability that the spinner will land on red can be written in all of these ways: 1 out of 2, $\frac{1}{2}$, 0.5, and 50%.

The probability that the spinner will land on green can be written in the same ways.

In many cases, the probability of an event will be greater than $\frac{1}{2}$ or less than $\frac{1}{2}$.

Example What do you expect will happen if you spin this spinner many times?

The spinner is divided into 4 sections that have the same shape and size. Three of the sections are green, and only one section is red. So the spinner should land on green about 3 times as often as it lands on red.

If you spin the spinner many times, it will land on red about 1 out of 4 spins, or about $\frac{1}{4}$ of the time. And it will land on green about 3 out of 4 spins, or about $\frac{3}{4}$ of the time.

The probability of landing on green can be written in all of these ways: 3 out of 4, $\frac{3}{4}$, 0.75, or 75%.

Sometimes you use an experiment to find a probability.

Example If you toss a tack, it can land with the point up or down. What is the chance that it will land point up?

You can do an experiment to find the chance of landing point up. Toss a large number of tacks and see how many land point up.

Suppose you toss 100 tacks and 70 land point up. The fraction landing point up is $\frac{70}{100}$.

So you can estimate the probability for a tack landing point up in any of these ways: 70 out of 100, $\frac{70}{100}$, 0.70, or 70%.

You can sometimes solve probability problems by making a list of the possible results.

Example Play this game. Put 2 red blocks and 3 blue blocks in a bag. Take one block out without looking. What is the chance of taking a red block?

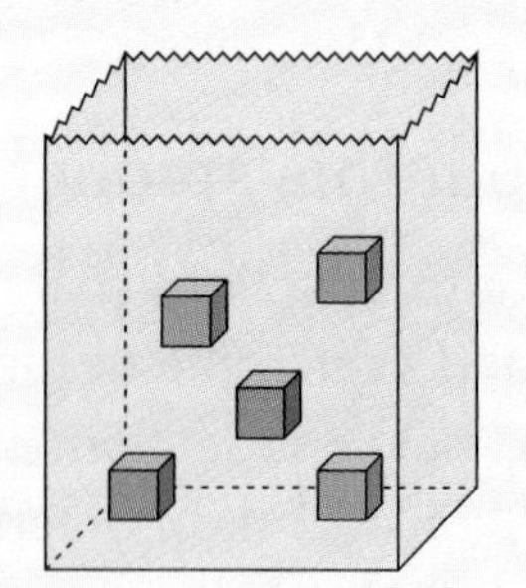

Make a list of the blocks and their colors:

block 1	block 2	block 3	block 4	block 5
red	red	blue	blue	blue

The blocks are the same, except for their color. So each block should have the same chance of being taken from the bag. Each block should have a 1 out of 5 chance of being taken.

Block 1 and block 2 are the two red blocks. You should take block 1 about 1 out of 5 times. And you should take block 2 about 1 out of 5 times.

So, for every 5 times you play this game, you should take a red block about 2 times. The probability of taking a red block can be written as 2 out of 5, or $\frac{2}{5}$.

Geometry

Points and Line Segments

In mathematics we study numbers. We also study shapes such as triangles, circles, and pyramids. The study of shapes is called **geometry.**

The simplest shape is a **point.** A point is a location in space. You often make a dot with a pencil to show where a point is. Name the point with a capital letter.

Here is a picture of 3 points. The letter names make it easy to talk about the points. For example, point A is closer to point B than it is to point P. And point B is closer to point A than it is to point P.

A B

P

A **line segment** is 2 points and the straight path between them. You can use any tool with a straight edge to draw the path between two points.

- The two points are called the **endpoints** of the line segment.
- The line segment is the shortest path between the endpoints.

The symbol for a line segment is a raised bar $\overline{}$. The bar is written above the letters that name the endpoints of the segment. The following line segment can be written as $\overline{AB}$ or as $\overline{BA}$.

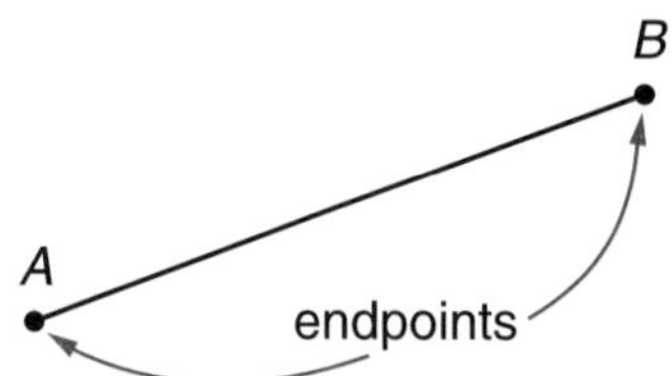

Rays and Lines

A **ray** is a straight path that has a starting point and goes on forever in *one* direction. You can draw a line segment with 1 arrowhead to stand for a ray.

Point R is the **endpoint** of this ray. The symbol for a ray is a raised bar with 1 arrowhead $\overrightarrow{}$. The ray shown here can be written $\overrightarrow{RA}$. The endpoint R is listed first. The second letter names some other point on the ray.

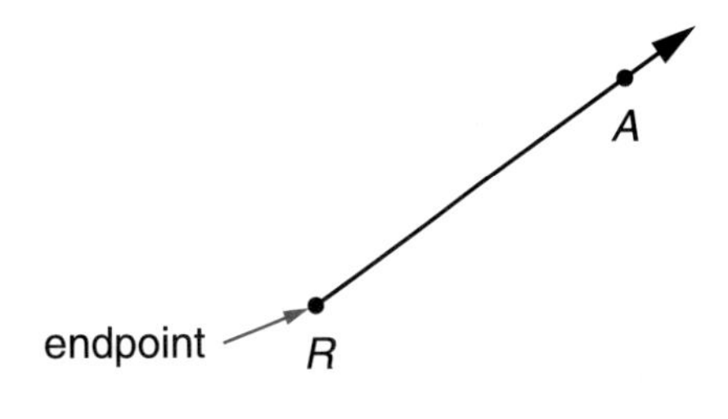

A **line** is a straight path that goes on forever in *both* directions. You can draw a line segment with 2 arrowheads to stand for a line. The symbol for a line is a raised bar with 2 arrowheads $\overleftrightarrow{}$.

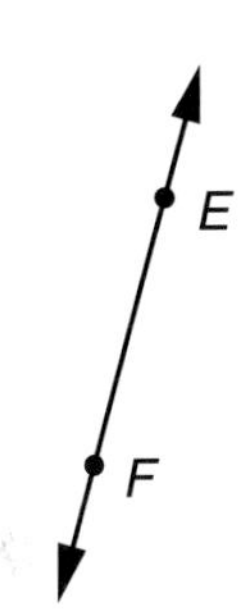

You can name a line by listing 2 points on the line. Then write the symbol for a line above the letters. The line here is written as $\overleftrightarrow{FE}$ or as $\overleftrightarrow{EF}$.

Example Write all the names for this line.

Points D, O, and G are all on the line. Use any 2 points to write the name of the line.

$\overleftrightarrow{DO}$ or $\overleftrightarrow{OD}$ or $\overleftrightarrow{DG}$ or $\overleftrightarrow{GD}$ or $\overleftrightarrow{OG}$ or $\overleftrightarrow{GO}$

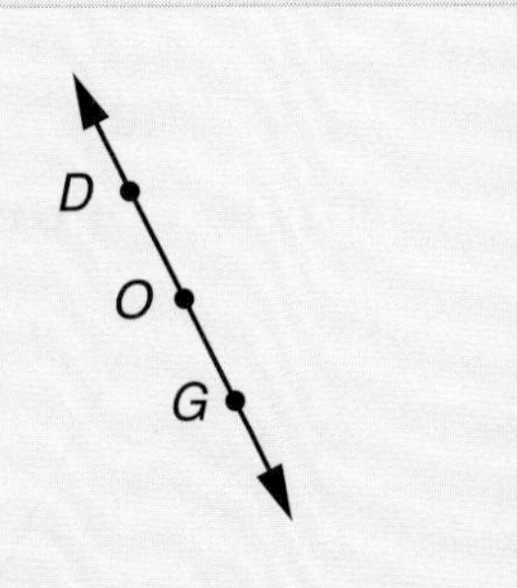

Angles

An **angle** is formed by 2 rays or 2 line segments that share the same endpoint.

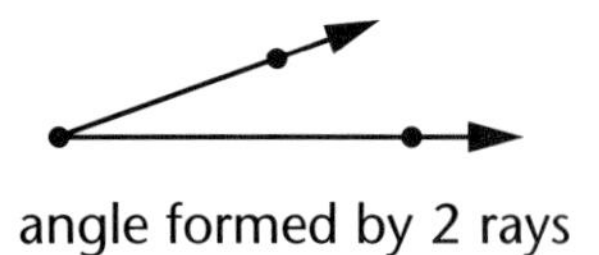

angle formed by 2 rays

angle formed by 2 segments

The endpoint where the rays or line segments meet is called the **vertex** of the angle. The rays or segments are called the **sides** of the angle.

∠ is the symbol for an angle. This is angle *T*, or ∠*T*.

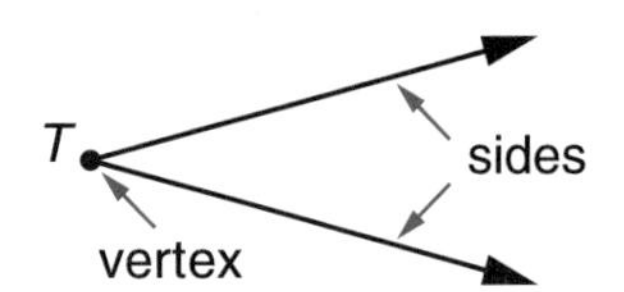

Angles can be measured with an angle measurer (protractor). Angles are measured in degrees. A **right angle** measures 90° (90 degrees). Its sides form a square corner. You often draw a small corner symbol inside the angle to show that it is a right angle.

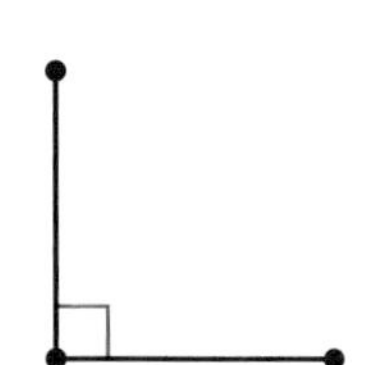

Example The small curved arrow in each picture shows which angle opening should be measured.

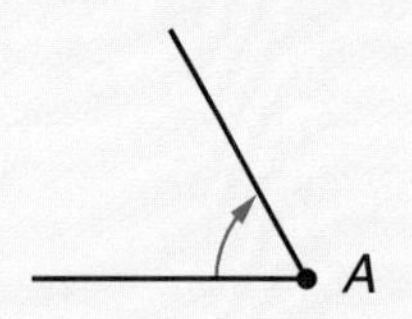

Measure of ∠*A* is 60°.

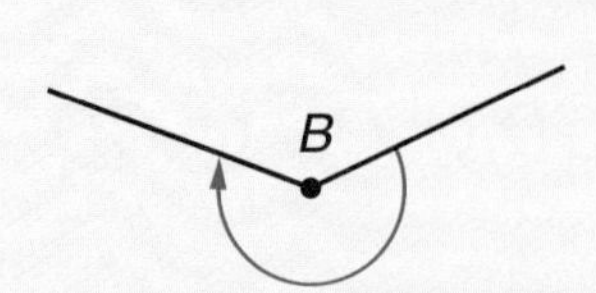

Measure of ∠*B* is 225°.

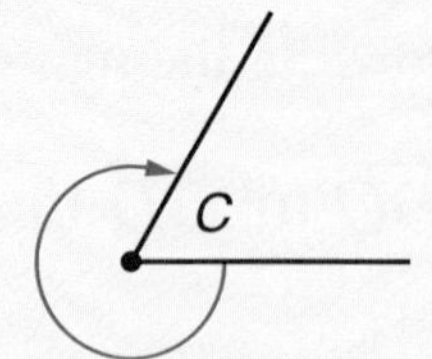

Measure of ∠*C* is 300°.

Parallel Lines and Segments

Parallel lines are lines that never meet and are always the same distance apart. Imagine a railroad track that goes on forever. The two rails are parallel. The rails never meet or cross. The rails are always the same distance apart (about 4 ft 8 in.).

Aerial photo of a railroad yard

Parallel line segments are segments that are parts of lines that are parallel. A section of railroad track has two rail segments that are parallel.

The symbol for *parallel* is a pair of vertical lines ||.

If lines or segments cross or meet each other, they **intersect.**

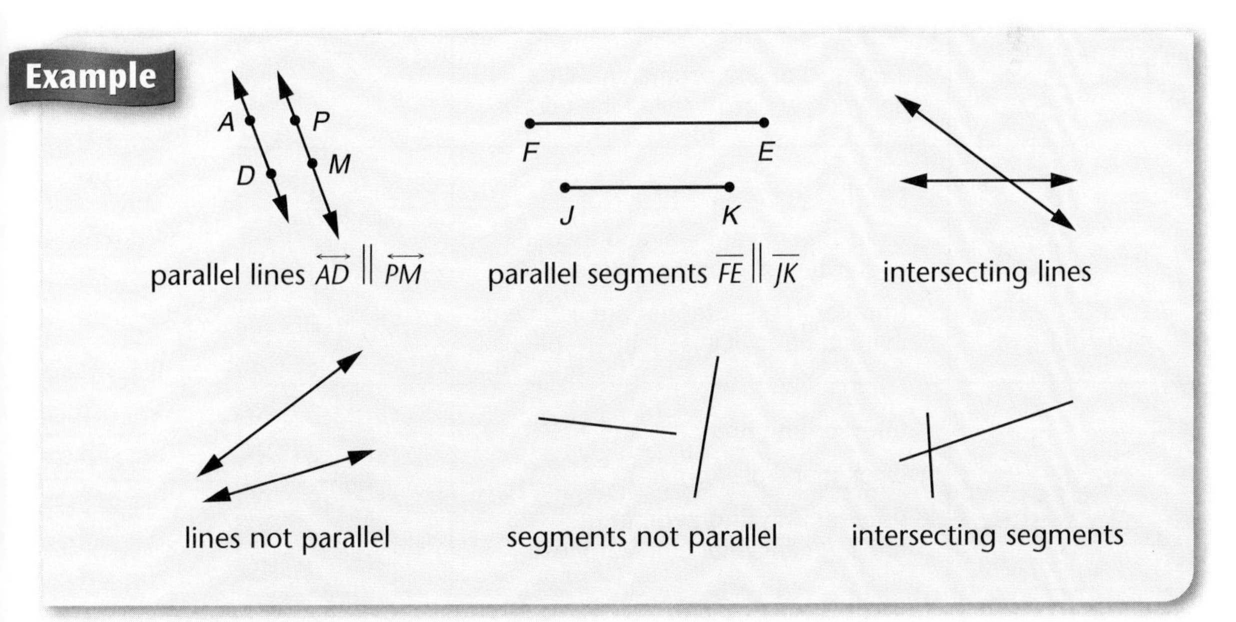

Line Segments, Rays, Lines, and Angles

Figure	Name and Description	Symbol
• A	**point** *A* A location in space	*A*
E, F, endpoints	**line segment** *EF* or *FE* A straight path between 2 points called its **endpoints**	$\overline{EF}$ or $\overline{FE}$
N, M, endpoint	**ray** *MN* A straight path that goes on forever in one direction from an **endpoint**	$\overrightarrow{MN}$
R, P	**line** *PR* or **line** *RP* A straight path that goes on forever in both directions	$\overleftrightarrow{PR}$ or $\overleftrightarrow{RP}$
vertex, T	**angle** *T* Two rays or line segments with a common endpoint called the **vertex**	$\angle T$
B, A, S, R	**parallel lines** *AB* and *RS* Lines that never meet and that are everywhere the same distance apart **Parallel line segments** are segments that are parts of lines that are parallel.	$\overleftrightarrow{AB} \parallel \overleftrightarrow{RS}$ $\overline{AB} \parallel \overline{RS}$
R, E, D, S	**intersecting lines** *DE* and *RS* Lines that cross or meet **Intersecting line segments** are segments that cross or meet.	none

line

angle

parallel lines

intersecting lines

Polygons

A **polygon** is a flat, 2-dimensional figure made up of 3 or more line segments called **sides.**

- The sides of a polygon are connected end to end and make one closed path.
- The sides of a polygon do not cross.

Each endpoint where two sides meet is called a **vertex.** The plural of vertex is *vertices.*

Figures That Are Polygons

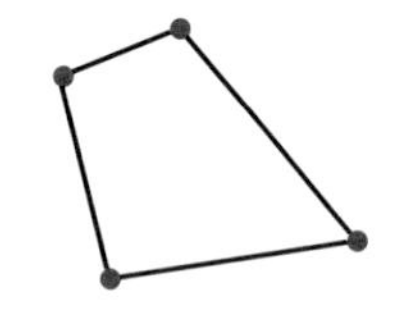

4 sides, 4 vertices

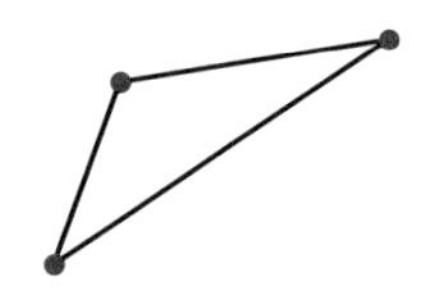

3 sides, 3 vertices

7 sides, 7 vertices

Figures That Are NOT Polygons

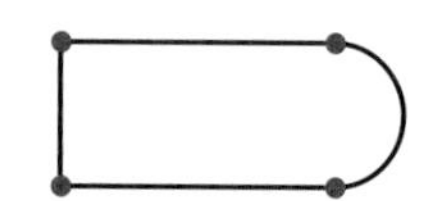

All sides of a polygon must be line segments. Curved lines are not line segments.

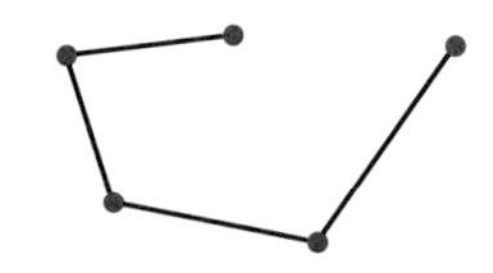

The sides of a polygon must form a closed path.

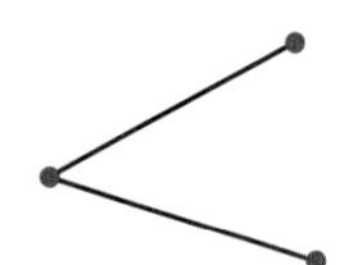

A polygon must have at least 3 sides.

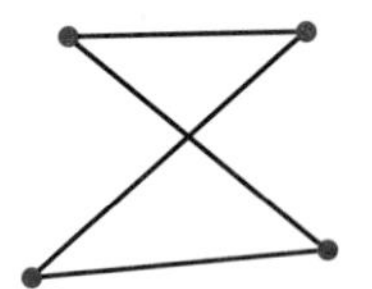

The sides of a polygon must not cross.

Polygons are named after the number of their sides. The prefix for a name tells the number of sides.

triangle

quadrangle or quadrilateral

pentagon

hexagon

octagon

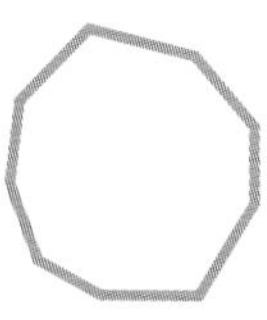

nonagon

heptagon

Prefixes

tri-	3
quad-	4
penta-	5
hexa-	6
hepta-	7
octa-	8
nona-	9
deca-	10
dodeca-	12

Did You Know?

The flag of Nepal is the only flag in the world with 5 sides. All other flags have 4 sides. The flag of Switzerland has a white cross with an edge that is a dodecagon (12 sides).

flag of Nepal

flag of Switzerland

Check Your Understanding

1. Name the polygon.
   - **a.** 6 sides
   - **b.** 4 sides
   - **c.** 10 sides
   - **d.** 8 sides
   - **e.** 12 sides
2. Draw a pentagon whose sides are not all the same length.

Check your answers on page 339.

A **regular polygon** is a polygon whose sides all have the same length and whose angles are all the same size.

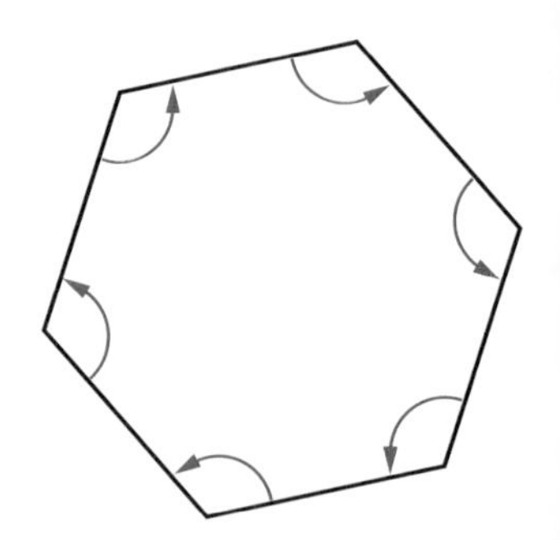

The hexagon shown here is a regular hexagon. All six sides are the same length. All six angles are the same size.

Some regular polygons have special names.

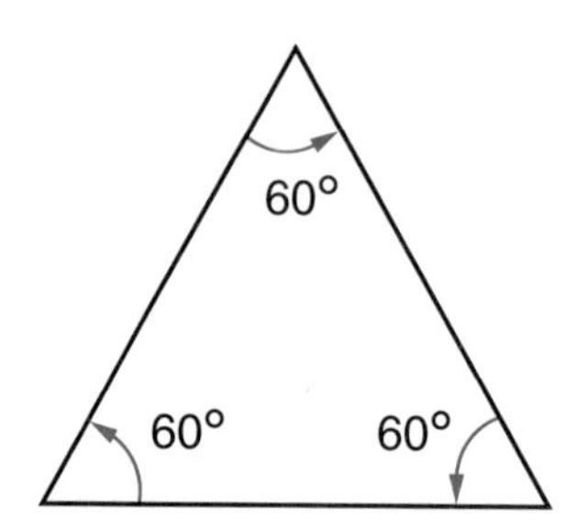

A regular triangle is called an **equilateral triangle.** It has 3 sides that are the same length. It has 3 angles that each measure 60°.

A design using equilateral triangles

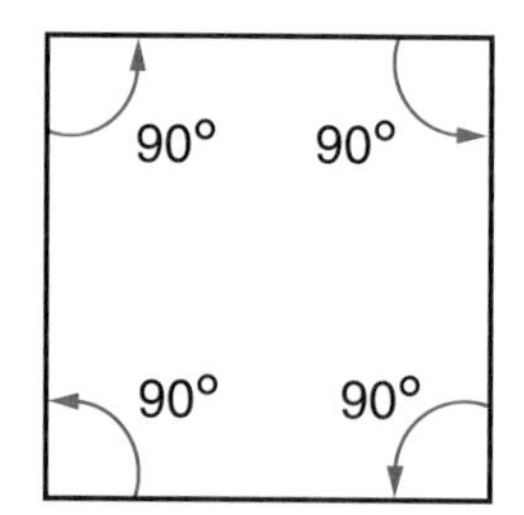

A **square** is a regular quadrangle (quadrilateral). It has 4 sides that are the same length. It has 4 angles that each measure 90°.

Laying a square tile floor

The Pentagon in Washington, D.C. has a ground area of 1,263,240 square feet. It has a larger ground area than any other office building in the world. The outside walls of the Pentagon have the shape of a regular pentagon. The inside walls of the Pentagon also have the shape of a regular pentagon. The 5 angles of a regular pentagon each measure 108°.

A stop sign has the shape of a regular octagon. It has 8 sides that are the same length. It has 8 angles that each measure 135°.

Triangles

Triangles are the simplest type of polygon. The prefix "tri-" means *three*. All triangles have 3 sides, 3 vertices, and 3 angles.

Example Name the parts of the triangle below.

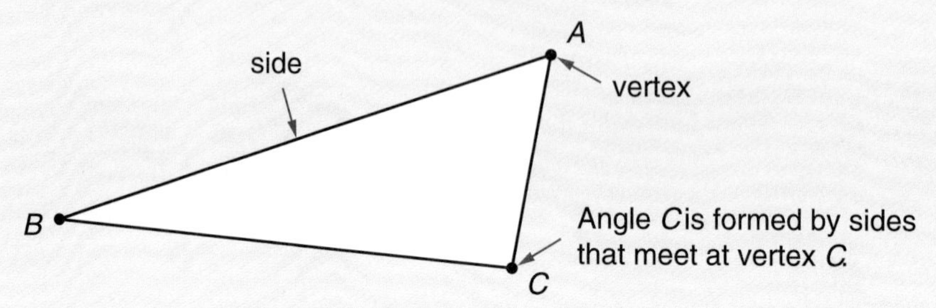

The sides are $\overline{BC}$, $\overline{BA}$, and $\overline{CA}$. The sides are the line segments that form the triangle.

The vertices are the points *B*, *C*, and *A*. Each endpoint where two sides meet is called a *vertex*.

The angles are $\angle B$, $\angle C$, and $\angle A$. An angle is formed by the two sides that meet at a vertex. For example, $\angle B$ is formed by $\overline{BC}$ and $\overline{BA}$.

Triangles have 3-letter names. You name a triangle by listing the letters for each vertex in order. The triangle in the example above has 6 possible names:

triangle *BCA*, *BAC*, *CAB*, *CBA*, *ABC*, or *ACB*.

Triangles have many different sizes and shapes.
Two special types of triangles have been given names.

Equilateral Triangles

An **equilateral triangle** is a triangle with all 3 sides the same length. All equilateral triangles have the same shape.

Right Triangles

A **right triangle** is a triangle with 1 right angle (square corner). Right triangles can have many different shapes.

Other triangles are shown below. None of these is an equilateral triangle. None is a right triangle.

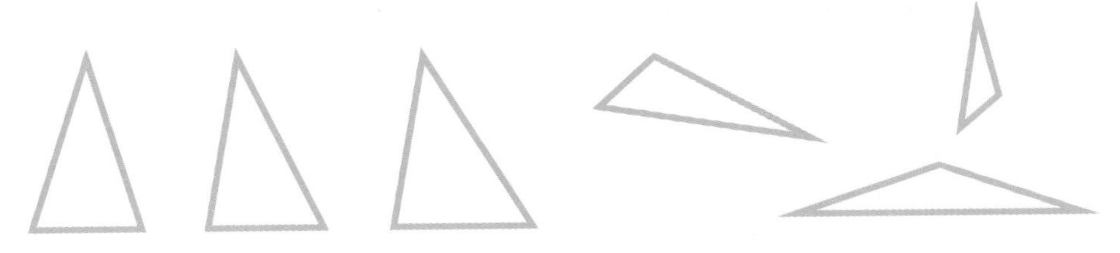

Quadrangles

A **quadrangle** is a polygon that has 4 sides. Another name for *quadrangle* is **quadrilateral.** The prefix "quad-" means *four*. All quadrangles have 4 sides, 4 vertices, and 4 angles.

Did You Know?

A quadricycle is a vehicle similar to the bicycle and tricycle but having 4 wheels.

Example Name the parts of the quadrangle.

The sides are $\overline{RS}$, $\overline{ST}$, $\overline{TU}$, and $\overline{UR}$.
The vertices are R, S, T, and U.
The angles are $\angle R$, $\angle S$, $\angle T$, and $\angle U$.

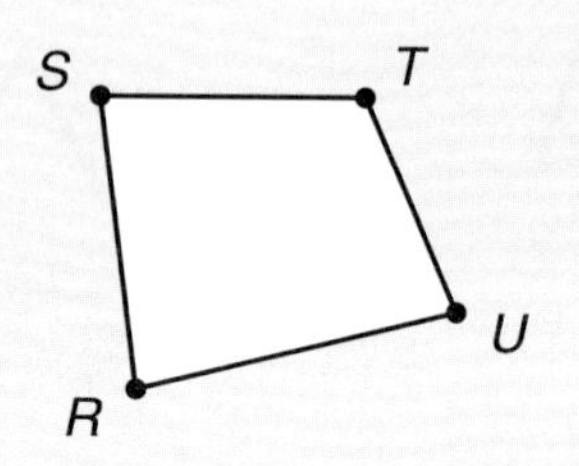

Some quadrangles have 2 pairs of parallel sides. These quadrangles are called **parallelograms.**

Reminder: Two sides are parallel if they are parts of lines that are parallel (never cross).

Figures That Are Parallelograms

Opposite sides are parallel in each figure.

Figures That Are NOT Parallelograms

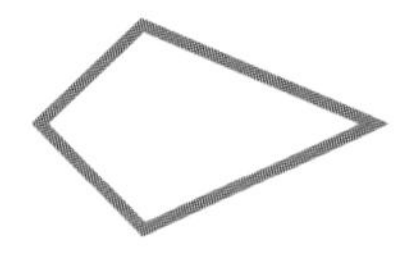

no parallel sides

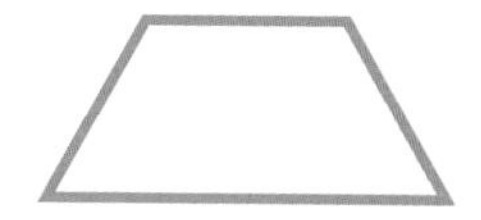

only 1 pair of parallel sides

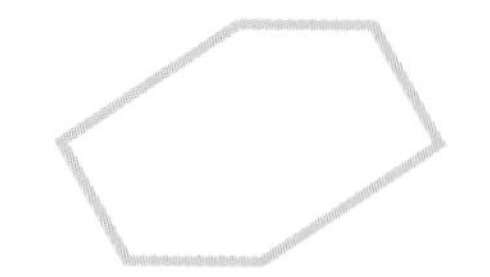

3 pairs of parallel sides, but a parallelogram must have 4 sides

Some quadrangles have special names.
Some of them are parallelograms.
Others are not parallelograms.

Quadrangles That Are Parallelograms

	Rectangles are parallelograms. They have 4 right angles (square corners). The sides of a rectangle do not all have to be the same length.
	Rhombuses are parallelograms. Their 4 sides are all the same length.
	Squares are parallelograms. They have 4 right angles (square corners). Their 4 sides are all the same length. All squares are rectangles. All squares are rhombuses.

Quadrangles That Are NOT Parallelograms

	Trapezoids have exactly 1 pair of parallel sides. Their 4 sides can all be different lengths.
	Kites are 4-sided polygons with 2 pairs of equal sides. The equal sides are next to each other. Their 4 sides cannot all be the same length. A rhombus is not a kite because all 4 sides of the rhombus are the same length.
	others Any polygon with 4 sides that is not a parallelogram, a trapezoid, or a kite

Circles

A **circle** is a curved line that forms a closed path. All of the points on a circle are the same distance from the **center of the circle.**

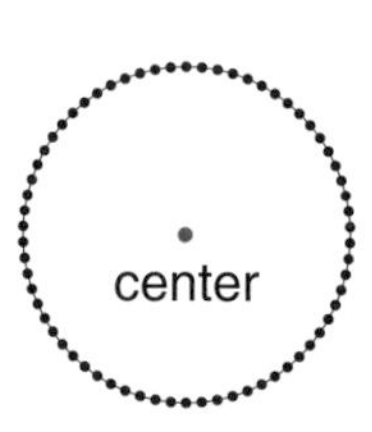

Circles are flat 2-dimensional figures. We can draw circles on a sheet of paper.

All circles have the same shape, but they do not all have the same size. The size of a circle is the distance across the circle through its center. This distance is called the **diameter of the circle.**

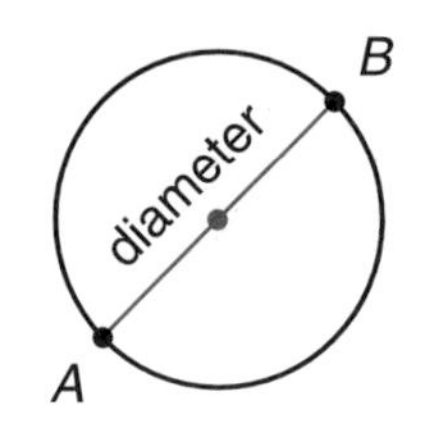

Line segment *AB* is a diameter of the circle. $\overline{AB}$ has a length of $\frac{7}{8}$ in., so the diameter of the circle is $\frac{7}{8}$ in.

The word *diameter* has another meaning. Any line segment that goes through the center of the circle and has endpoints on the circle is called a **diameter of the circle.** The figure at the right shows both uses of the word *diameter*.

Concentric circles are circles that have the same center but different diameters.

Examples of concentric circles

Example Many pizzas have the shape of a circle. We often order a pizza by saying the diameter we want.

A "12-inch pizza" means a pizza with a 12-inch diameter.

A "16-inch pizza" means a pizza with a 16-inch diameter.

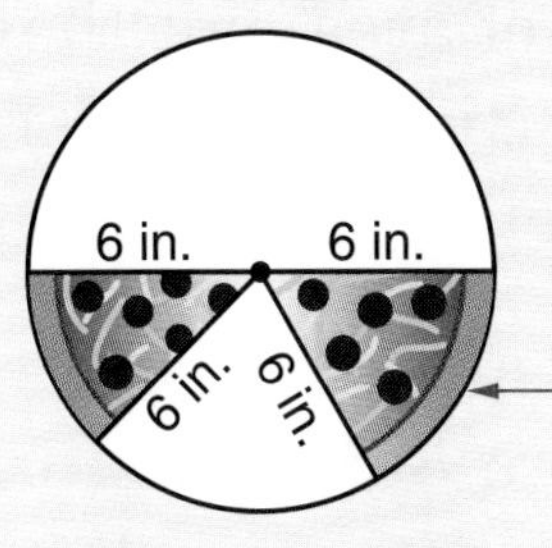

A 12-inch pizza

Check Your Understanding

Use your ruler to measure the diameter of each circle.

Measure to the nearest quarter inch.

1. 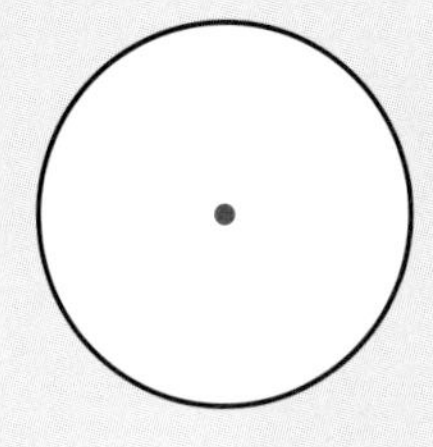

2. 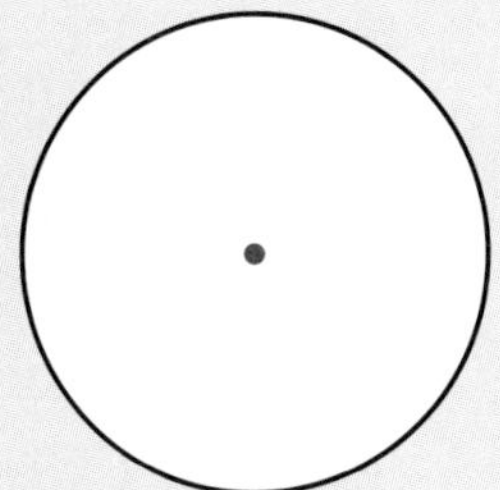

3. 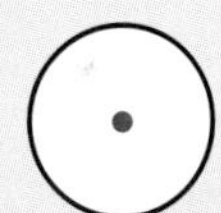

Measure to the nearest centimeter.

4. 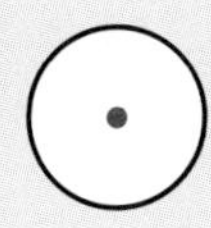

5. 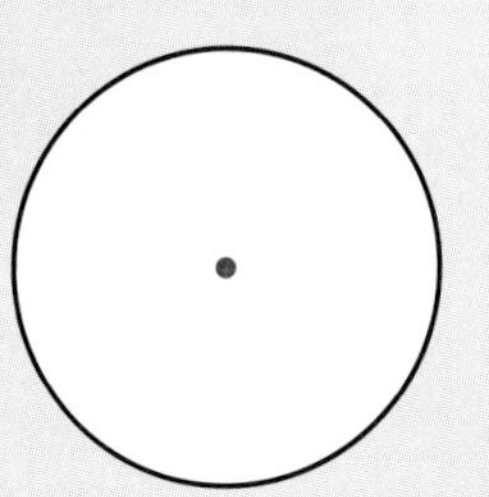

6. 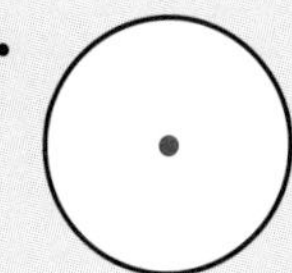

Check your answers on page 339.

Solids

Triangles, quadrangles, and circles are flat shapes. They take up a certain amount of area, but they do not take up space. They are flat, **2-dimensional** figures. We can draw the figures on a sheet of paper.

Solid objects that take up space are such things as boxes, books, and chairs. They are **3-dimensional** objects. Some solids, such as rocks and animals, do not have regular shapes. Other solids have shapes that are easy to describe using geometric words.

The **surfaces** on the outside of a solid may be flat or curved or both. A **flat surface** of a solid is called a **face.**

A cube has 6 faces.

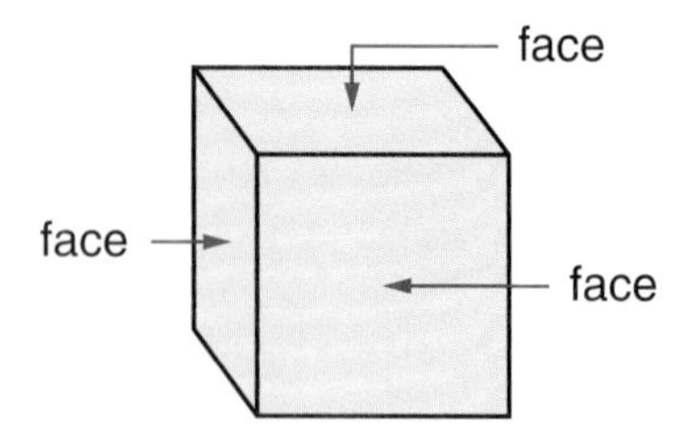

There are 3 other faces that cannot be seen in this picture.

This box is shaped like a cube.

A cylinder has 2 faces and 1 curved surface.

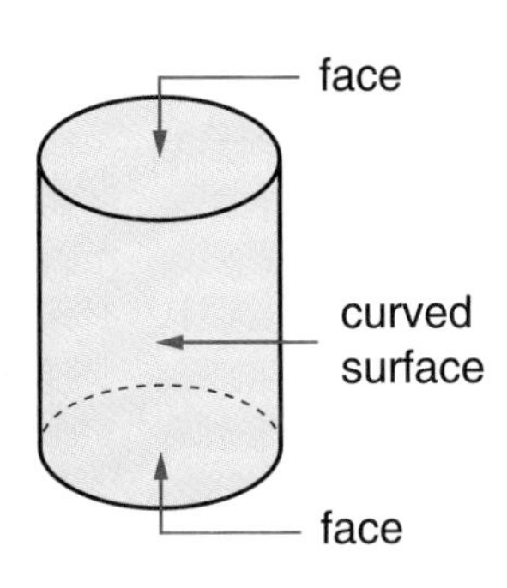

The dashed line shows a hidden edge.

The table top is a short cylinder with large faces. The center pole is a tall cylinder with small faces.

A cone has 1 face and 1 curved surface.

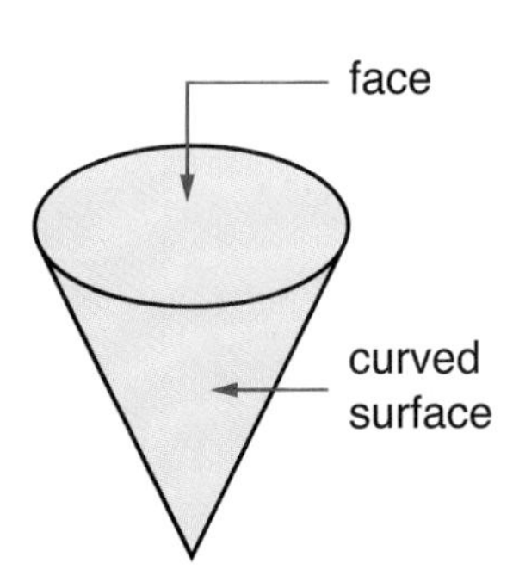

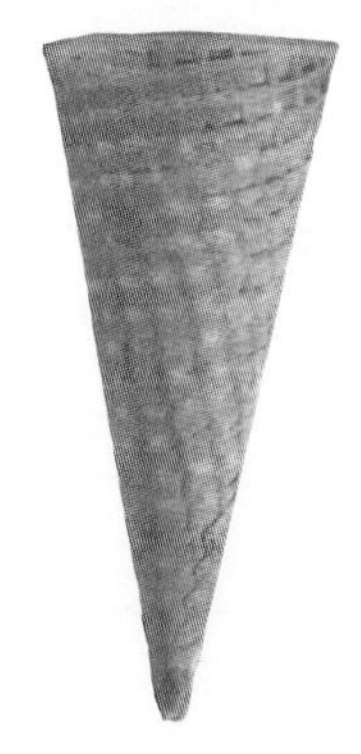

This ice cream cone is a good model of a cone. But keep in mind that a cone is closed. It has one face that acts like a cover.

Did You Know?

Buildings shaped like cylinders and cones are not often seen. Here are some examples.

The Rotunda is an office building in Birmingham, England.

This structure is covered with glazed ceramic. It is the entrance to a parking garage in Valencia, Spain.

The surfaces of a solid meet one another. They form curves or line segments. These curves or line segments are the **edges** of the solid.

Example Identify the edges of a cube, a cylinder, and a cone.

A cube has 12 edges. A cylinder has 2 edges. A cone has 1 edge.

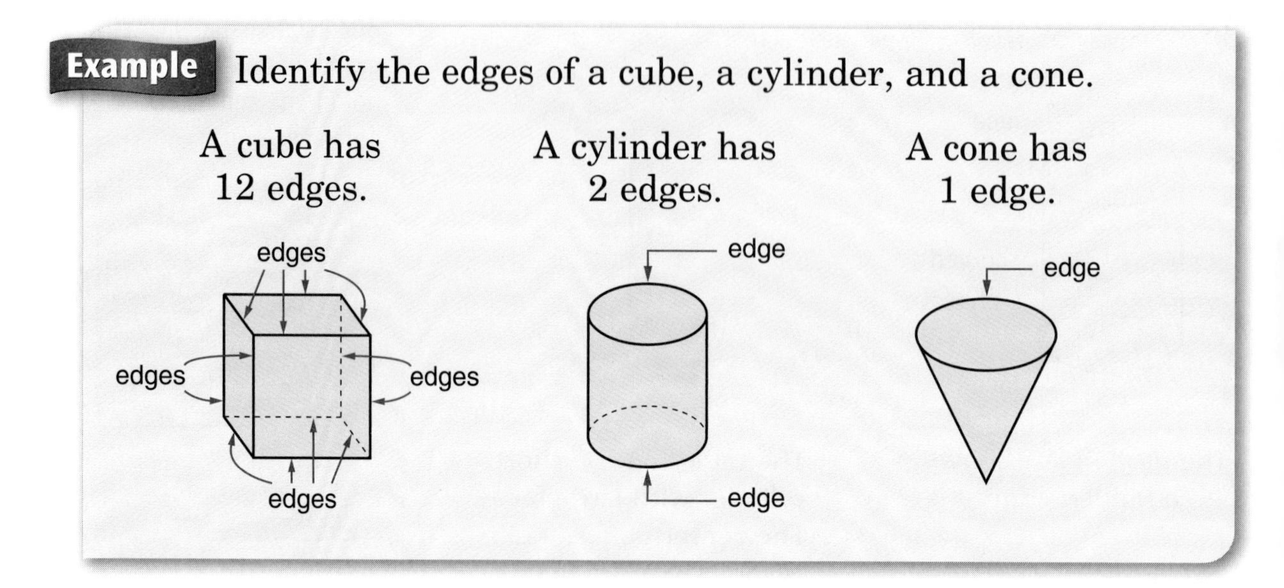

A corner of a solid is called a **vertex.**

Example Identify the vertices of a cube, a cylinder, and a cone.

A cube has 8 vertices. A cylinder has 0 vertices. A cone has 1 vertex.

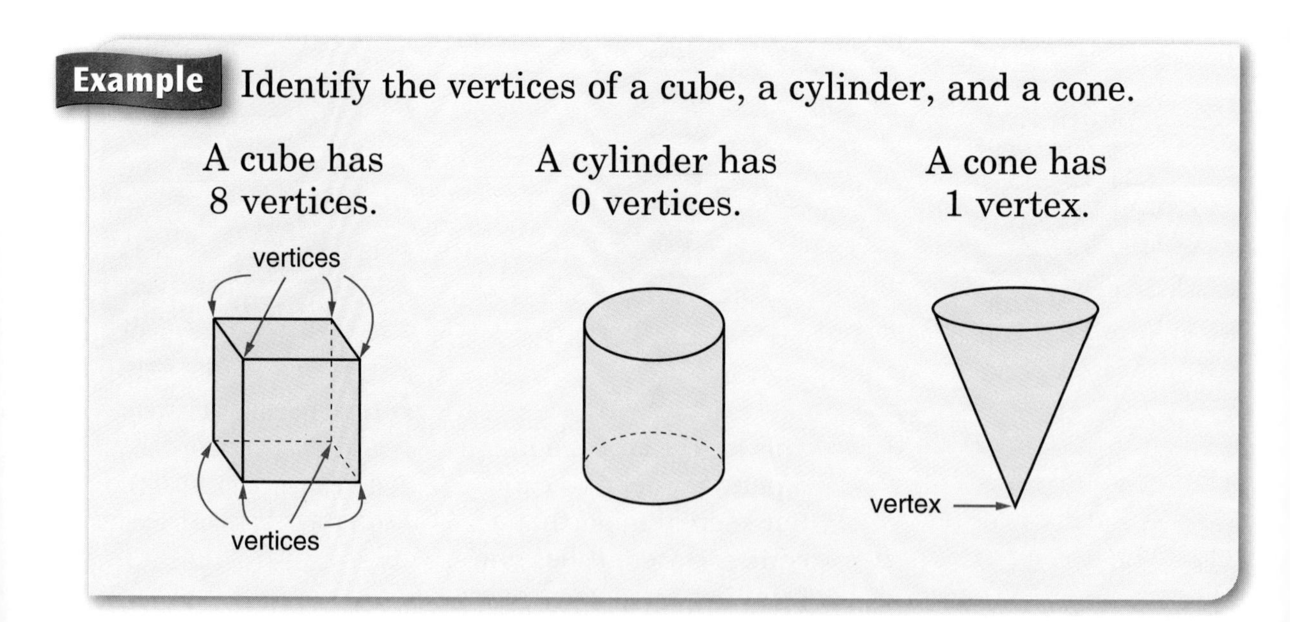

Polyhedrons

A **polyhedron** is a solid whose surfaces are all flat and formed by polygons. It does not have any curved surfaces. The faces of a **regular polyhedron** are all formed by copies of one regular polygon that have the same size.

Three important groups of polyhedrons are shown below. These are **pyramids, prisms,** and **regular polyhedrons.** Many polyhedrons do not belong to any of these groups.

Pyramids

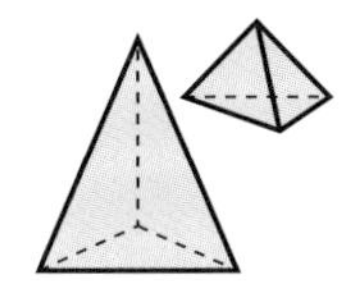

triangular pyramids

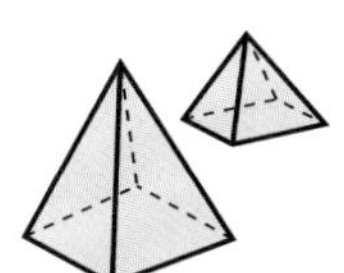

rectangular pyramids

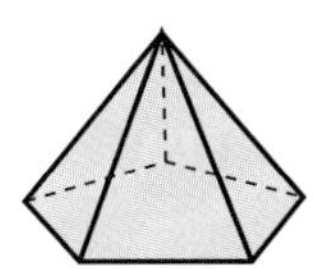

pentagonal pyramid

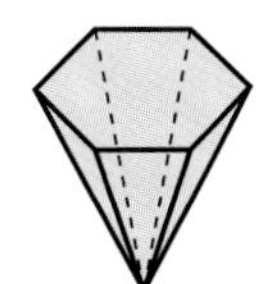

hexagonal pyramid

Prisms

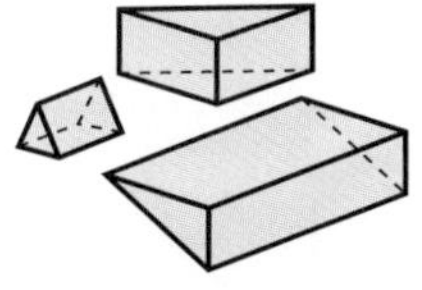

triangular prisms

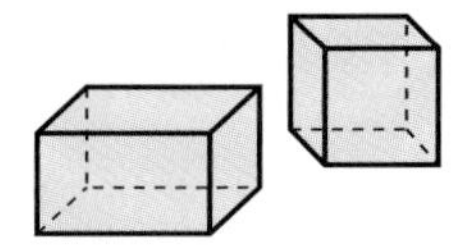

rectangular prisms

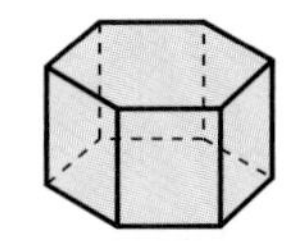

hexagonal prism

Regular Polyhedrons

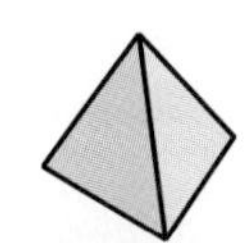

regular tetrahedron (pyramid) (4 faces)

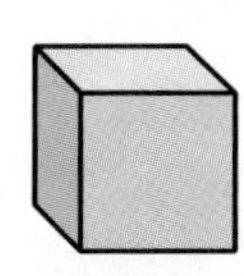

cube (prism) (6 faces)

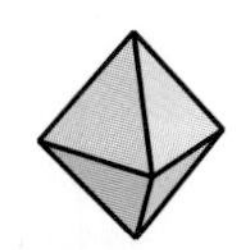

regular octahedron (8 faces)

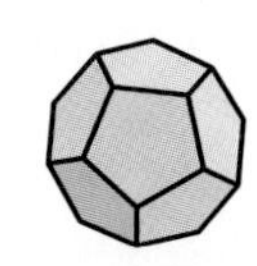

regular dodecahedron (12 faces)

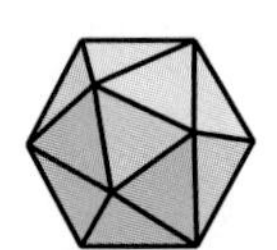

regular icosahedron (20 faces)

Pyramids

All of the solids below are **pyramids.**

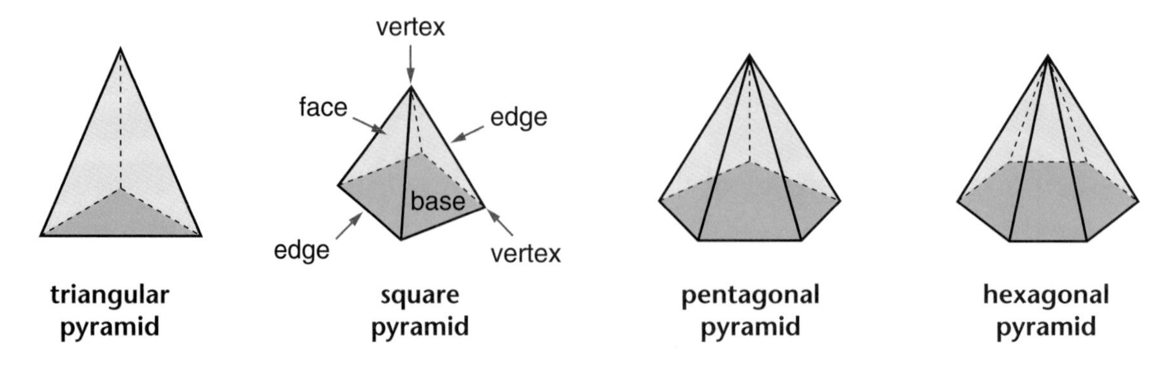

Pyramids have flat surfaces called **faces.**

The *shaded* face of each pyramid above is called the **base** of the pyramid. The faces that are not the base all have the shape of a triangle. The faces that are not the base all come together at one vertex.

The shape of the base is used to name a pyramid. If the base has the shape of a triangle, the pyramid is called a **triangular pyramid.** If the base has the shape of a square, the pyramid is called a **square pyramid.**

The pyramids of Egypt have square bases. They are called square pyramids.

Prisms

All of the solids below are **prisms.**

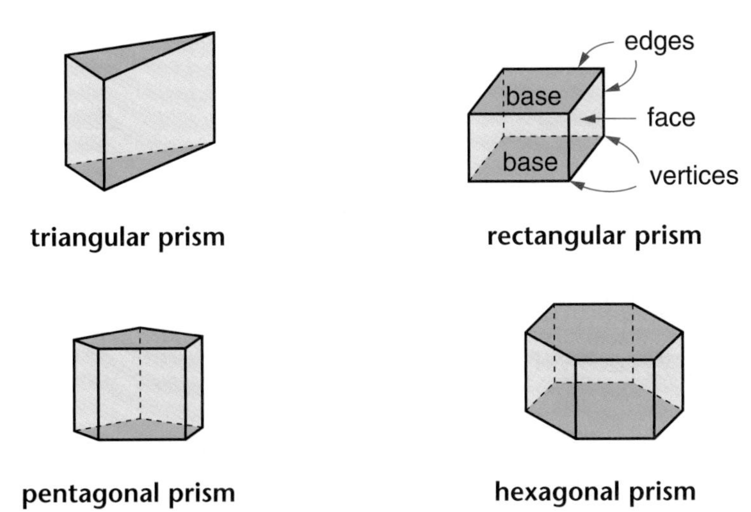

triangular prism

rectangular prism

pentagonal prism

hexagonal prism

Prisms have flat surfaces called **faces.**

The two *shaded* faces of each prism above are called the **bases** of the prism.

- Both bases have the same size and shape.
- The bases are parallel. This means that the bases are everywhere the same distance apart.
- The faces that connect the bases are shaped like rectangles or parallelograms.

The shape of the bases is used to name a prism. If the bases have the shape of a triangle, the prism is called a **triangular prism.** If the bases have the shape of a rectangle, the prism is called a **rectangular prism.**

Cylinders and Cones

A **cylinder** has two flat surfaces that are connected by a curved surface. Soup cans and paper towel rolls are shaped like cylinders.

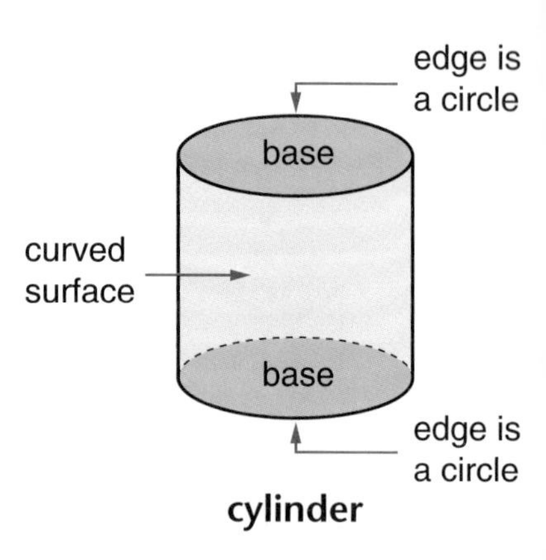

cylinder

The flat surfaces are called **bases.**

- The 2 bases are shaped like circles. These circles are the same size.
- The 2 bases are parallel. This means that the bases are everywhere the same distance apart.

These candles are shaped like cylinders.

Another solid with a curved surface is a **cone.** Many ice cream cones and some paper cups are shaped like cones.

A cone has 1 flat surface that is shaped like a circle. This is the **base** of the cone. The other surface of the cone is a curved surface. The curved surface wraps around the base. It ends at a point called the **apex** of the cone.

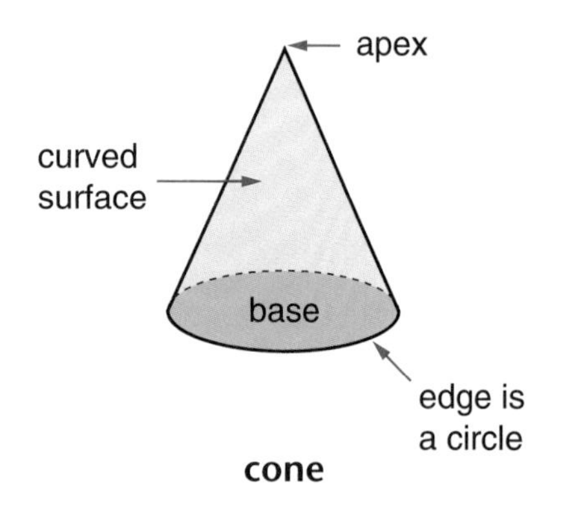

cone

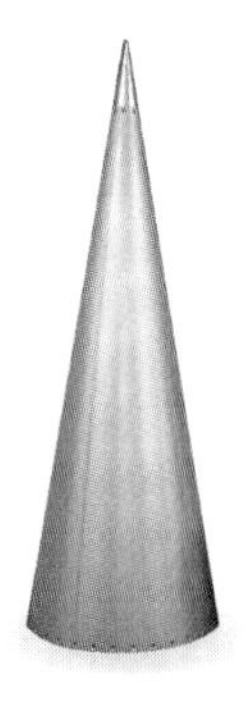

This nose cone is part of a rocket.

This party hat is shaped like a cone.

Spheres

A **sphere** is a solid with a curved surface that is shaped like a ball or a globe. All of the points on the sphere's surface are the same distance from the **center of the sphere.**

Spheres are 3-dimensional objects. They take up space. All spheres have the same shape. But not all spheres have the same size.

Did You Know?

Baked clay marbles shaped like spheres date back to 3000 B.C.

The size of a sphere is the distance across the sphere and through its center. This distance is called the **diameter of the sphere.**

The segment *RS* passes through the center of the sphere. The length of this segment is the diameter of the sphere. The segment *RS* is also called a diameter of the sphere.

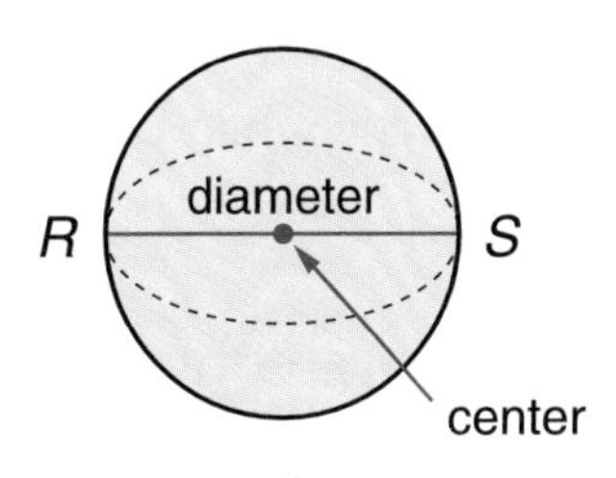

Example Earth is shaped very nearly like a sphere. The diameter of Earth is about 8,000 miles. The distance from Earth's surface to the center of Earth is about 4,000 miles. Every point on Earth's surface is about 4,000 miles from the center of Earth.

Layers inside Earth

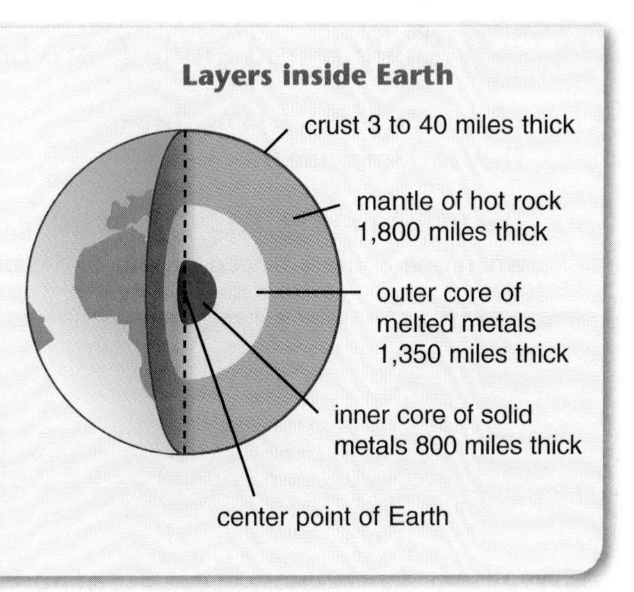

Congruent Figures

Sometimes figures have the same shape and same size. We say that these figures are **congruent.** Figures are congruent if they match exactly when one figure is placed on top of the other.

Line segments are congruent if they have the same length.

Example Line segments AB and CD are both 3 cm long.

Both segments have the same shape.
Both segments have the same length.

The line segments are congruent. They match exactly when one segment is placed on top of the other.

Angles are congruent if they have the same degree measure.

Example Angle E and angle F are both right angles.

$\angle E$ and $\angle F$ have the same shape.
They each measure 90°.

The angles are congruent. They match exactly when one angle is placed on top of the other.

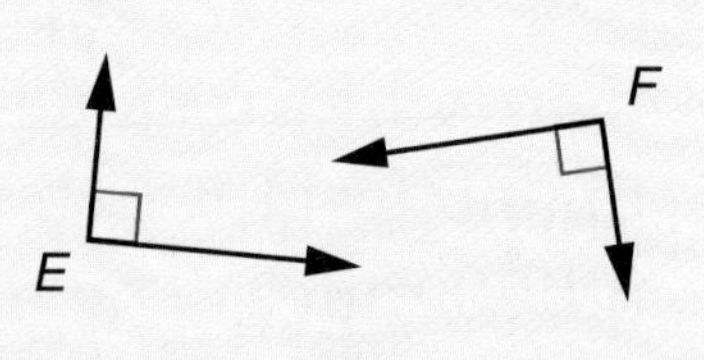

Circles are congruent if their diameters are the same length.

Example The circles here have $\frac{1}{2}$-inch diameters.

All 3 circles have the same shape.
All 3 circles have the same size.

The circles are congruent. They match exactly when one is placed on top of the other.

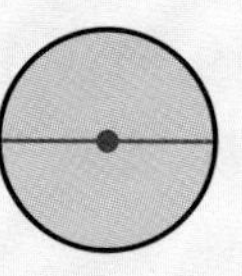
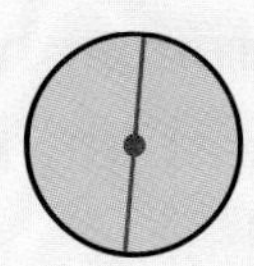
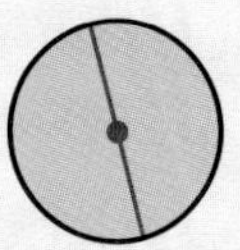

If we use a copy machine to copy a figure, the original and the copy are congruent.

Example A copy machine was used to copy the pentagon *RSTUV*.

If we cut out the copy, it will match exactly when placed on top of the original.

The sides will match exactly. All the angles will match exactly.

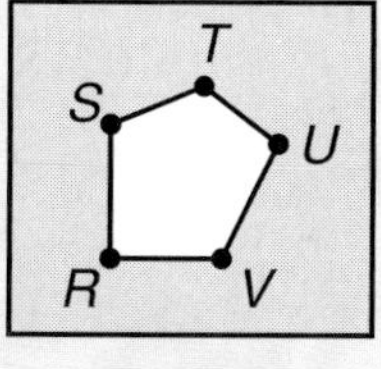

original

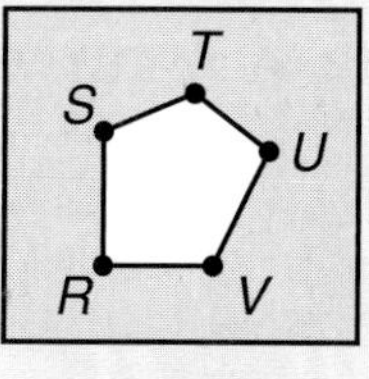

copy

Check Your Understanding

Which one of the following triangles is NOT congruent to the other three?

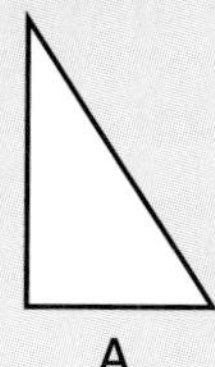
A

B

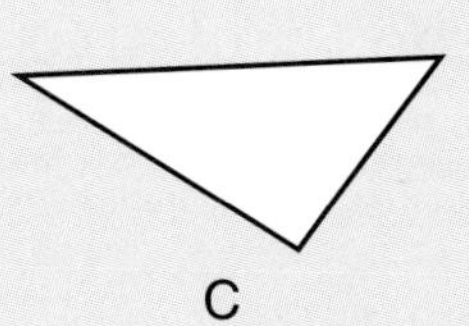
C

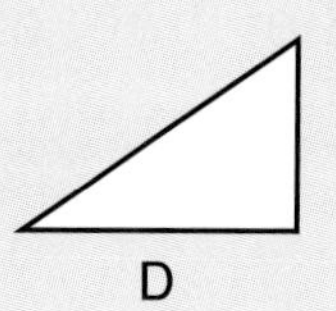
D

Check your answer on page 339.

Line Symmetry

Look at this photograph of a butterfly. A dashed line is drawn through it. The line divides the photograph into two parts. Both parts look alike. Each is a mirror image of the other.

The figure is **symmetric about a line.** The dashed line is called a **line of symmetry** for the figure.

An easy way to find out whether a figure has line symmetry is to fold it in half. If the two halves match exactly, then the figure is symmetric. The fold line is the line of symmetry.

Example The letters T, V, E, and X are symmetric. The lines of symmetry are drawn below for each letter.

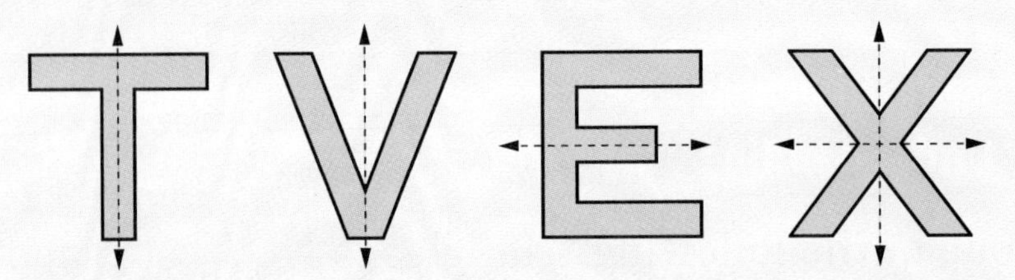

The letter X has two lines of symmetry. You can fold along either line, and the two halves will match exactly.

The figures below are all symmetric. The line of symmetry is drawn for each figure. If there is more than one line of symmetry, they are all drawn.

Figures That Are Symmetric about a Line

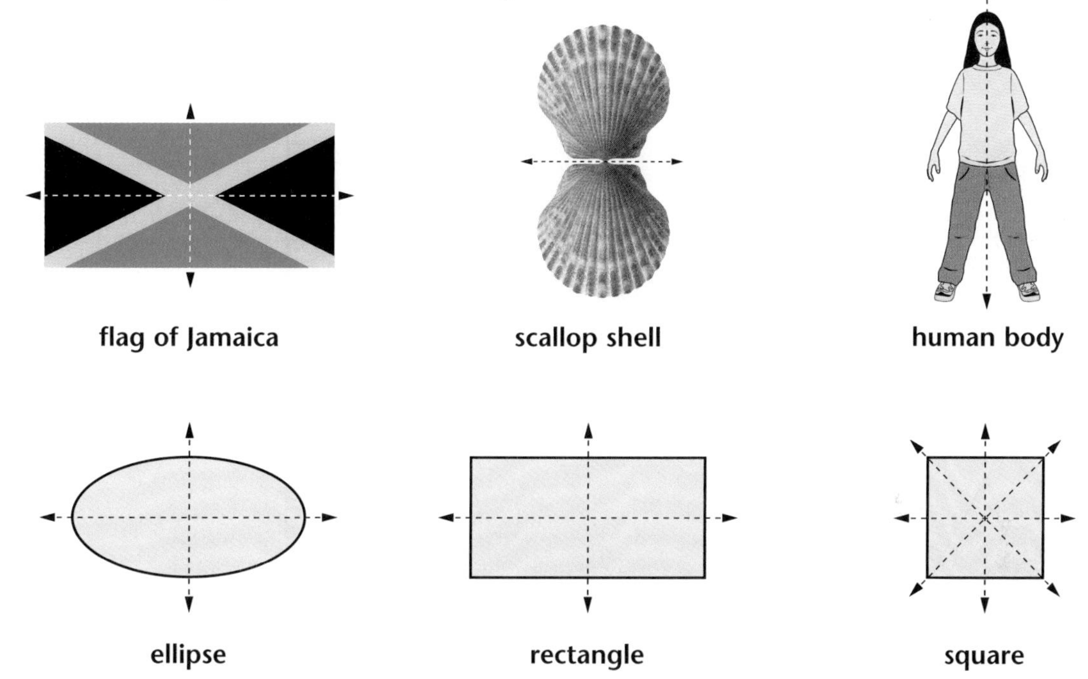

Check Your Understanding

1. Trace each pattern-block shape onto a sheet of paper. Draw all lines of symmetry for each shape.
2. How many lines of symmetry does a circle have?

Check your answers on page 340.

The Pattern-Block Template

There are 13 geometric figures on the Pattern-Block Template. All of the geometric figures are symmetric. Six of the figures are the same size as actual pattern blocks.

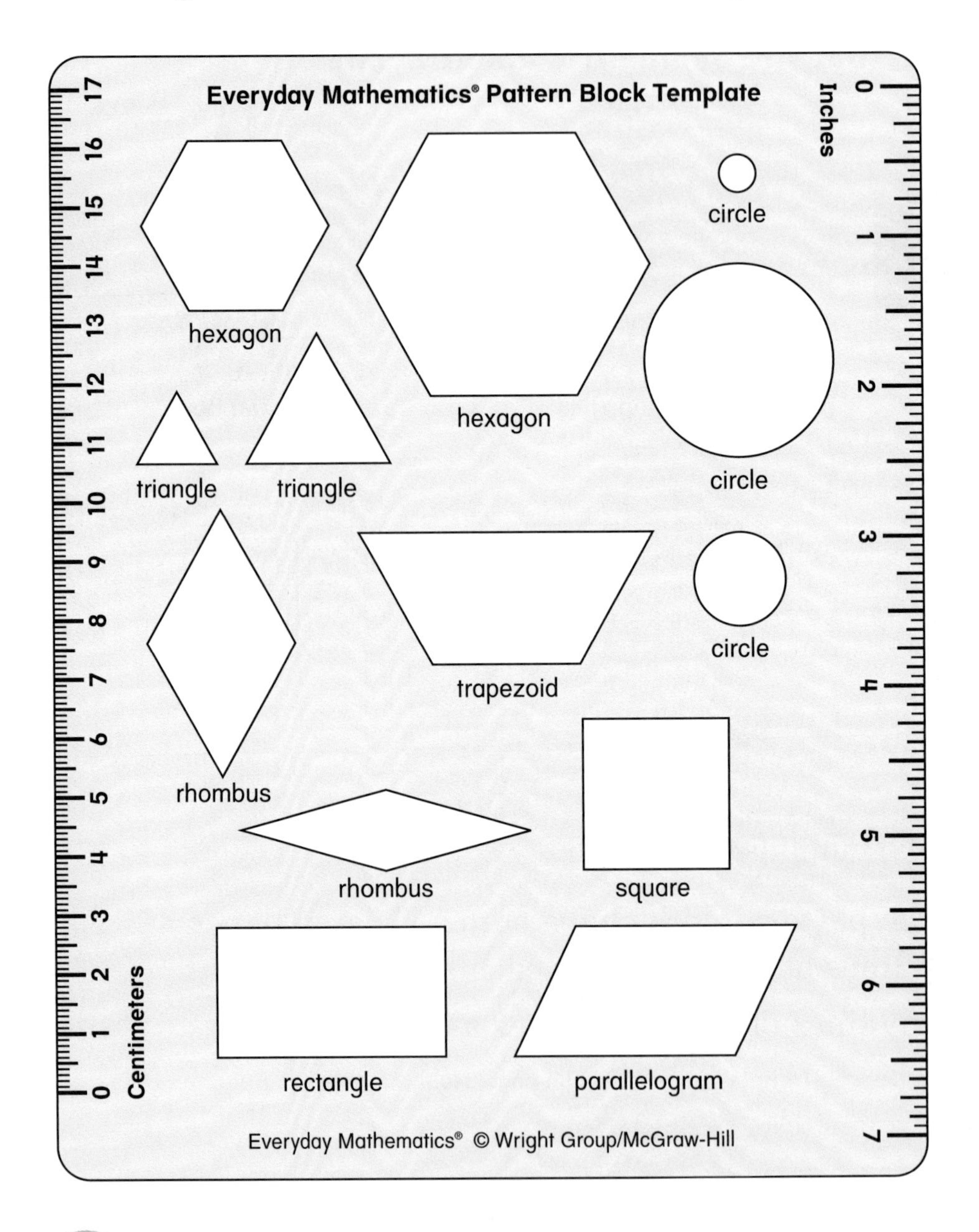

Mathematics ... Every Day

Geometry in Nature

Many interesting shapes in nature result from the way things grow or form. You can see some of these shapes with the naked eye. You can see others with special equipment, such as magnifying lenses, microscopes, or telescopes.

If you look at one of your fingertips with a powerful magnifying lens, you will see many curves. The curves of each unique fingerprint provide traction so you can lift things. ➤

Because they are so far away, the stars of the Big Dipper look like points when viewed from the earth. ⮟

Polygons, Circles, and Spheres

If you imagine a line from tip to tip on each arm of this sea star, you will make a pentagon. ➤

Honeycombs look like connected hexagons. These hexagons fit together with no space in between. They make a great place for bees to store honey. ⮟

⮝ Look closely at the shapes in this dried mud in Death Valley, California. Do you see shapes like quadrilaterals?

Mathematics ... Every Day

The circular leaves of this Victoria Regia water lily can grow up to two meters in diameter.

This is the spherical head of a dandelion that has gone to seed.

If you cut an onion in half, you will see what look like concentric circles.

Spirals

This photograph was taken from above the clouds of a hurricane. Do you see the spirals? ➤

The Meller's Chameleon coils its tail into a spiral. ➤

Sunflower seeds spiral in many different directions from the center. ➤

Mathematics ... Every Day

Three-Dimensional Shapes and Solids

When magnified using high-powered microscopes, you can see that these salt crystals are rectangular solids. ▼

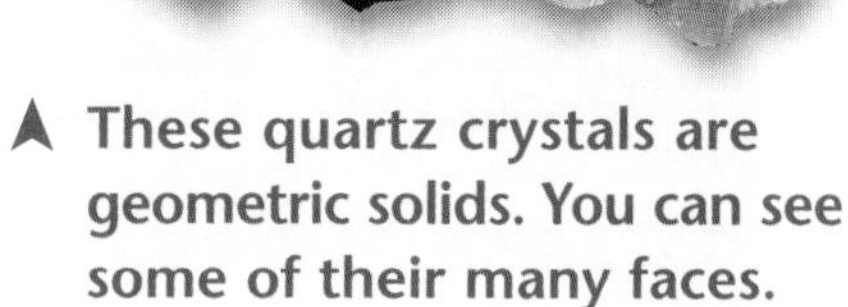

▲ These quartz crystals are geometric solids. You can see some of their many faces.

▲ These hexagonal rock columns of the Giant's Causeway in Ireland were shaped during volcanic activity.

The cone-shaped funnel cloud of a tornado descends to the ground. ▼

Symmetry

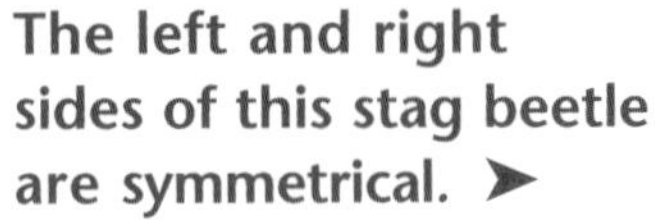

The left and right sides of this stag beetle are symmetrical. ➤

➤ The two halves of this artichoke are symmetrical.

Magnifying snowflakes shows that they are symmetrical. What shapes do you see in these snowflakes? ➤

➤ This is a close-up of an orchid flower. Do you see the symmetry between one side and the other?

Look around you. Where do you see geometry in nature?

Measurement

Measurement Before the Invention of Standard Units

People measured length and weight long before they had rulers and scales. In the past, people used parts of their bodies to measure lengths. Here are some units of length that were based on the human body.

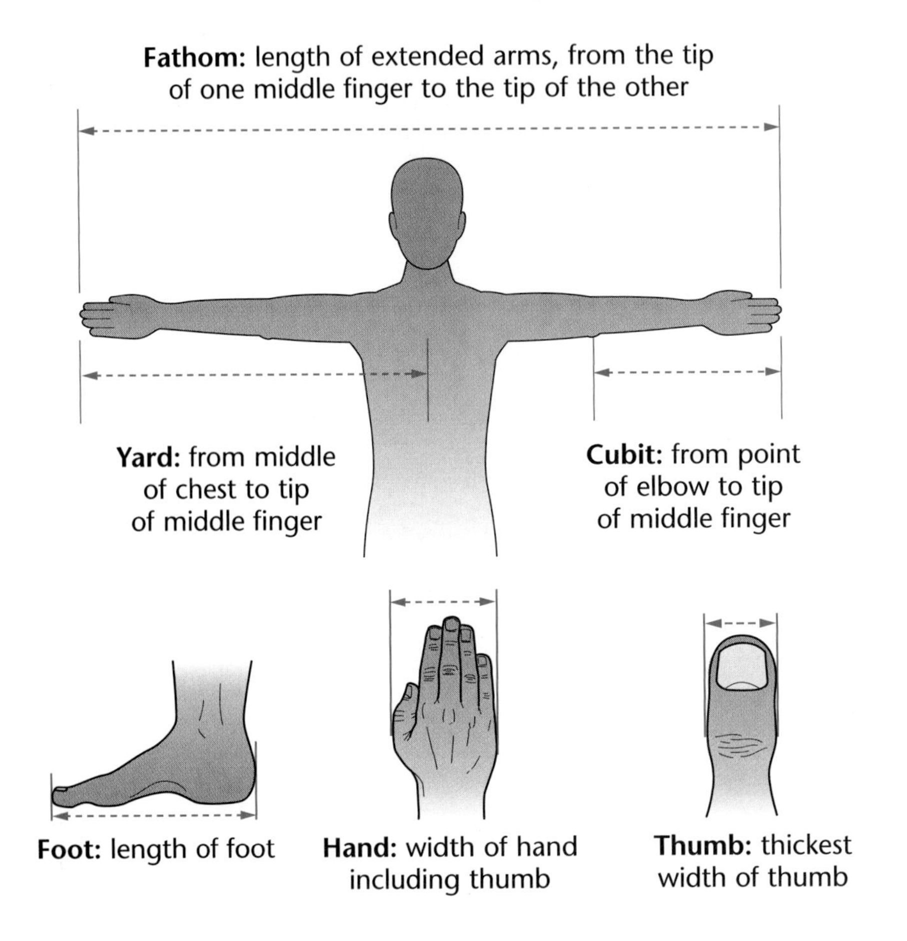

Look carefully. You will see that a fathom is about the same length as a person's height.

The problem with using body measures is that they are different for different people.

Using **standard units** of length solves this problem. The standard units never change. They are the same for everyone.

If two people measure the same object using standard units, their measurements will be the same or almost the same.

The Metric System

About 200 years ago, the **metric system** of measurement was developed. The metric system uses standard units for measuring length, weight, and temperature.

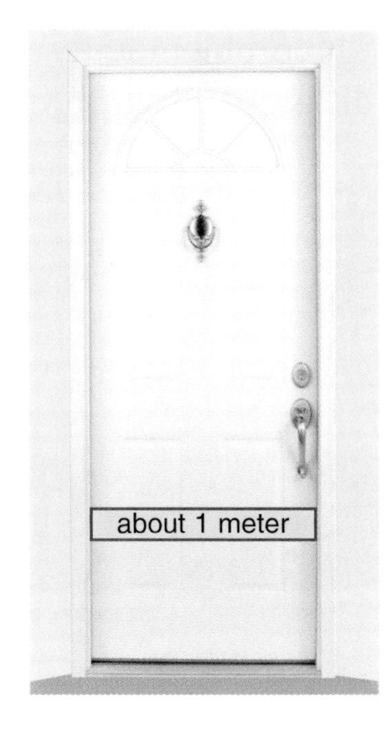

- The standard unit for length is the **meter.** The word *meter* is abbreviated **m.** A meter is about the length of a big step, or the width of a front door.
- The standard unit for weight is the **gram.** The word *gram* is abbreviated **g.** A dime weighs about 2 grams. A paper clip weighs about $\frac{1}{2}$ gram. So 2 paper clips weigh about 1 gram.
- The standard unit for temperature is the **Celsius degree** or **°C.** Water freezes at 0°C. Room temperature is usually about 20°C.

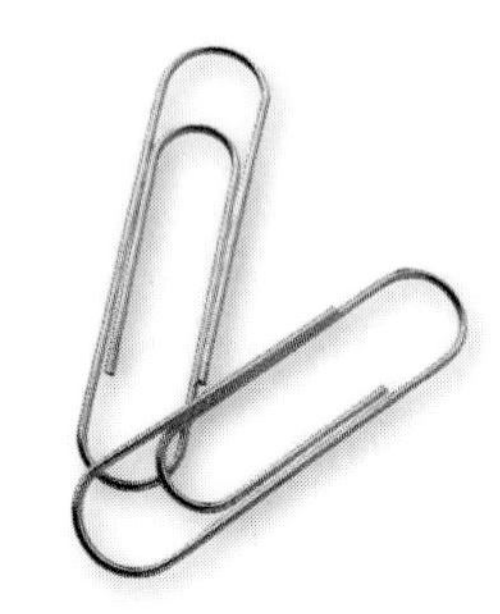

The metric system is used all over the world.

Scientists almost always measure using the metric system. Metric units are often used in sports such as track and field, ice skating, and swimming. Many food labels include metric measurements.

The metric system is easy to use because it is a decimal system. It is based on the numbers 10, 100, and 1,000.

Let's see what this means by looking at a **meterstick.** A meterstick is a ruler that is 1 meter long. There are probably metersticks in your classroom.

Example Part of a meterstick is shown below. It has been divided into smaller units.

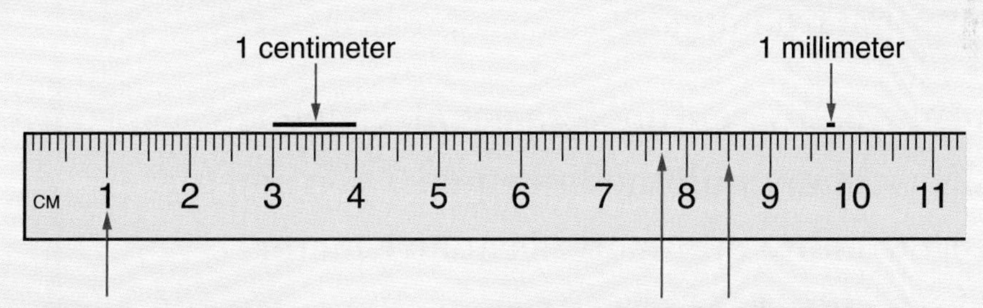

The meterstick is divided into 100 equal sections. The length of each section is called a **centimeter.** There are 100 centimeters in 1 meter.

Each centimeter is divided into 10 equal sections. The length of each small section is called a **millimeter.**

There are 10 millimeters in 1 centimeter.
There are 1,000 millimeters in 1 meter.

The United States is the only large country in which the metric system is not used for everyday measurements. Often the **U.S. customary system** is used instead. The U.S. customary system uses standard units such as the **inch, foot, yard,** and **pound.**

The U.S. customary system is not based on the numbers 10, 100, and 1,000. This makes it more difficult to use than the metric system. For example, to change inches to yards, you must know that 36 inches equals 1 yard.

Check Your Understanding

1. **a.** How many centimeters equal 1 meter?
   b. How many millimeters equal 1 centimeter?
   c. How many millimeters equal 1 meter?
2. Which units below are in the metric system?

foot	millimeter	pound	inch
gram	meter	centimeter	yard

3. **a.** Draw a line segment that is 4 centimeters long.
   b. Draw another line segment that is 40 millimeters long.
   c. Which line segment is longer?

Check your answers on page 340.

Measuring Length in Centimeters and Millimeters

Length is the measure of a distance between two points. Length is usually measured with a ruler. The edges of your Pattern-Block Template are rulers. Tape measures, yardsticks, and metersticks are rulers that are used for measuring longer distances.

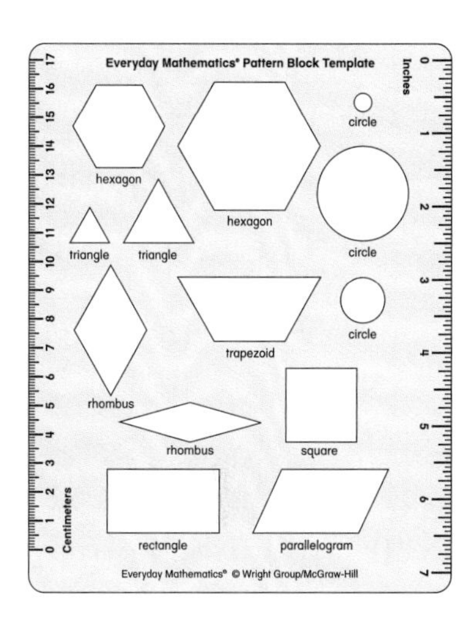

Rulers are often marked with **inches** on one edge and **centimeters** on the other edge. The side showing centimeters is called the **centimeter scale.** The side showing inches is called the **inch scale.**

Each centimeter is divided into 10 equal parts called **millimeters.** A millimeter is $\frac{1}{10}$ or 0.1 of a centimeter. The word *centimeter* is abbreviated **cm.** The word *millimeter* is abbreviated **mm.**

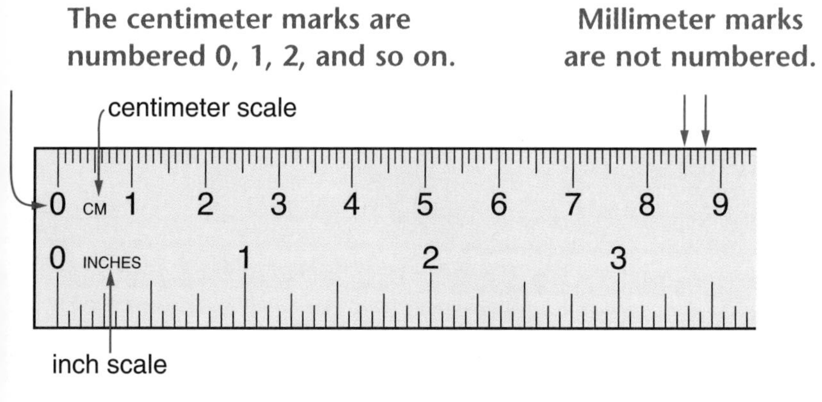

Did You Know?

In January 2003, Chinese scientists found the body of a four-winged dinosaur that was about 75 centimeters long from head to tail.

Example How long is the key?

Always line up one end of the object with the 0-mark on the ruler.

The other end of the key is at the 3-centimeter mark.

The key is 3 centimeters long.

We write this as 3 cm.

If the 0-mark is at the end of a ruler, then the number "0" may not be printed on the ruler. When this happens, line up the end of the object with the end of the ruler.

Example How many millimeters long is the arrow?

The 0-mark is at the end of the ruler. Line up the end of the object with the end of the ruler.

There are 40 millimeters from the end of the ruler to the 4 cm mark. The arrow tip is another 5 millimeters past the 4 cm mark.

So the arrow is 45 millimeters long.

We write this as 45 mm.

Example Find the length of the needle.

Use centimeters:

The end of the needle is 7 small spaces past the 2 cm mark. Each small space is $\frac{1}{10}$ of a centimeter.

So the needle is $2\frac{7}{10}$ (or 2.7) cm long.

Use millimeters:

There are 20 millimeters from the end of the ruler to the 2 cm mark. The needle tip is another 7 millimeters past the 2 cm mark.

So the needle is 27 mm long.

You can use a ruler to draw a line segment.

Example Draw a line segment that is 7.8 centimeters long.

Each centimeter equals 10 millimeters. So 0.1 cm equals 1 mm, and 0.8 cm equals 8 mm.

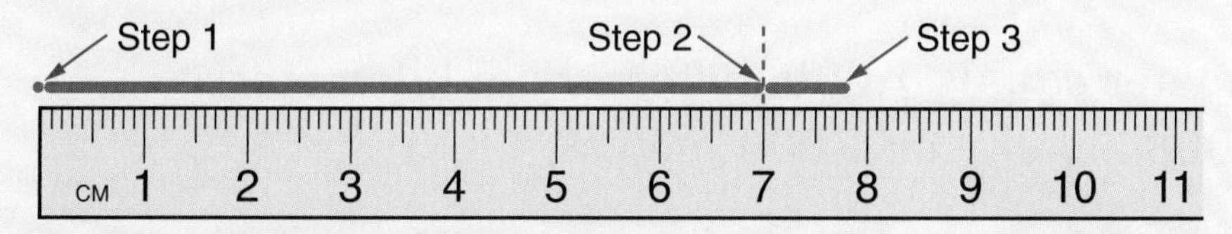

Step 1: Make a dot above the end of the ruler.

Step 2: Draw a line up to the 7 cm mark.

Step 3: Keep drawing until you have covered 8 more mm spaces.

Changing Units of Length in the Metric System

The basic unit of length in the metric system is the **meter.** We measure smaller lengths using **centimeters** and **millimeters.** We measure longer distances using **kilometers.** The table below shows how these different units of length compare.

Comparing Metric Units of Length		Abbreviations for Units of Length
1 cm = 10 mm	1 mm = $\frac{1}{10}$ cm	mm = millimeter
1 m = 100 cm	1 cm = $\frac{1}{100}$ m	cm = centimeter
1 m = 1,000 mm	1 mm = $\frac{1}{1,000}$ m	m = meter
1 km = 1,000 m	1 m = $\frac{1}{1,000}$ km	km = kilometer

You can use this reference table to change from one metric unit to another. All changes use the numbers 10, 100, or 1,000.

Example Jodi ran a 5-kilometer race. How many meters is 5 kilometers?

One kilometer equals 1,000 meters. So 5 kilometers is 5 × 1,000 meters, or 5,000 meters.

Example Mark is 150 centimeters tall. How many meters is 150 centimeters?

100 centimeters equals 1 meter, and 50 centimeters equals $\frac{1}{2}$ meter. So 150 cm equals 1 m + $\frac{1}{2}$ m, which is 1.5 meters.

Personal References for Metric Units of Length

Sometimes you may not have a ruler or meterstick handy. When this happens, you can estimate lengths by using the lengths of common objects and distances you know. Some examples are given below.

Personal References for Metric Units of Length	
About 1 millimeter	**About 1 centimeter**
Thickness of a dime Thickness of a pushpin point Thickness of a paper match	Thickness of your math journal Width of the head of a pushpin Thickness of a pattern block
About 1 meter	**About 1 kilometer**
Width of a door One big step (for an adult) Height of a kitchen counter	1,000 big steps (for an adult) Length of 10 football fields (including the end zones)

The point of the pushpin is about 1 millimeter thick.

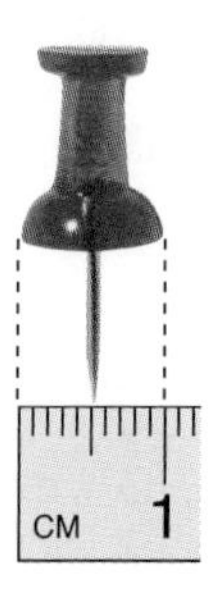

The head of the pushpin is about 1 centimeter wide.

The personal references for 1 meter can also be used for 1 yard. One meter is slightly longer than 39 inches. One yard equals 36 inches. So a meter is about 3 inches longer than a yard.

Did You Know?

At a height of about 979 meters, Angel Falls in Venezuela is the highest waterfall in the world.

meterstick

yardstick

Check Your Understanding

1. A 10K race is a 10-kilometer race. How many meters are in 10 kilometers?
2. **a.** 3 meters is how many centimeters?

   b. 3.5 meters is how many centimeters?
3. 40 mm is the same length as how many cm?
4. Would you measure the length of your bedroom in meters or millimeters?
5. **a.** 2,000 big steps by a man is about how many kilometers?

   b. About how many meters is that?
6. Ann made a stack of 25 dimes. About how many centimeters high is the stack?

Check your answers on page 340.

Measuring Length in Inches

Length is the measure of a distance between two points. In the U.S. customary system, a standard unit of length is the **inch.** The word *inch* is abbreviated **in.**

On rulers, inches are usually divided into halves, quarters (or fourths), eighths, and sixteenths. The marks to show fractions of an inch are usually different sizes.

This space is $\frac{1}{16}$ in. long.

This space is $\frac{4}{16}$ in. long or $\frac{1}{4}$ in. long.

These are the $\frac{1}{4}$-inch and $\frac{1}{2}$-inch marks between 3 and 4.

$3\frac{1}{4}$ $3\frac{2}{4}$ or $3\frac{1}{2}$ $3\frac{3}{4}$

0 INCHES 1 2 3 4

Example What is the length of each nail?

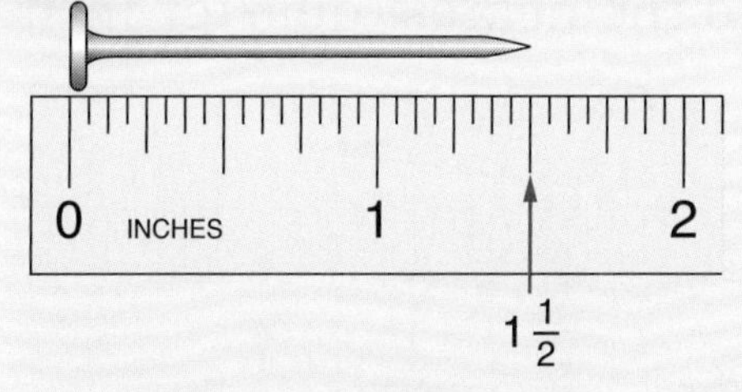

The end of the nail is at the $1\frac{1}{2}$-inch mark.

The nail is $1\frac{1}{2}$ inches long.

We can write this as $1\frac{1}{2}$ in.

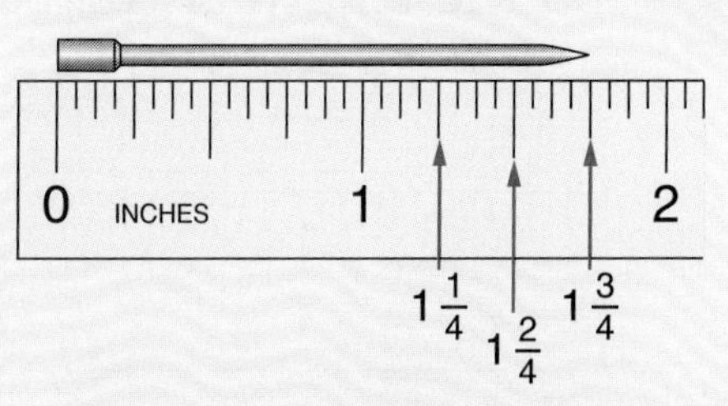

The $\frac{1}{4}$-inch marks between 1 and 2 are shown. The end of the nail is at the $1\frac{3}{4}$-inch mark.

It is $1\frac{3}{4}$ in. long.

Example What is the length of the eraser?

Always line up the end of the object with the 0-mark of the ruler.

If the 0-mark is at the end of a ruler, the number 0 may not be printed on that ruler.

There are 2 small spaces between the 2-inch mark and the end of the eraser. Each small space is $\frac{1}{16}$ inch long.

So the eraser is $2\frac{2}{16}$ inches long. Because $\frac{2}{16} = \frac{1}{8}$, the length can be written as $2\frac{2}{16}$ in., or $2\frac{1}{8}$ in.

There are times when you do not need an exact measurement. Measuring to "the nearest $\frac{1}{2}$ inch" or "the nearest $\frac{1}{4}$ inch" may be good enough.

Example Find the length of the pencil to the nearest quarter-inch.

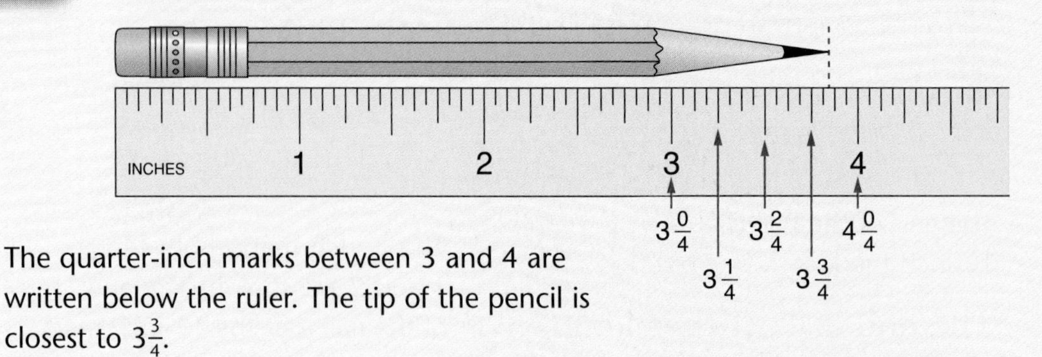

The quarter-inch marks between 3 and 4 are written below the ruler. The tip of the pencil is closest to $3\frac{3}{4}$.

The pencil is $3\frac{3}{4}$ inches long, to the nearest quarter-inch.

Check Your Understanding

1. Draw a line segment that is $2\frac{1}{2}$ inches long.

2. Measure the length of the crayon to the nearest quarter-inch.

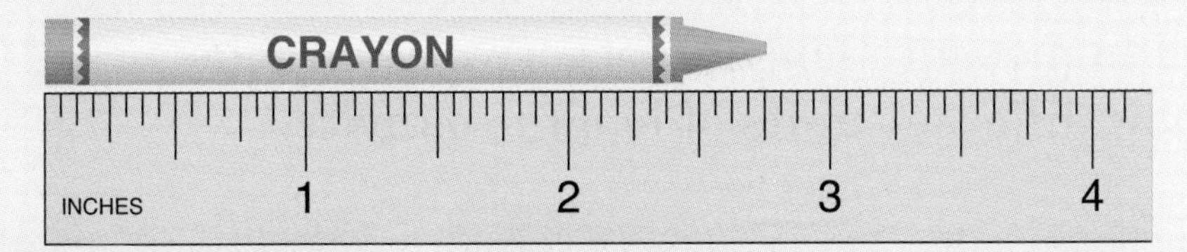

3. Name the measure shown by each letter.

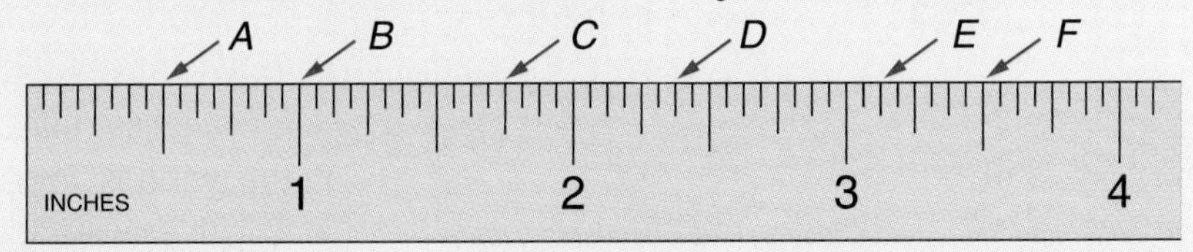

a. A is $\underline{\frac{1}{2}}$ in. b. B is _?_ in. c. C is _?_ in.

d. D is _?_ in. e. E is _?_ in. f. F is _?_ in.

Check your answers on page 340.

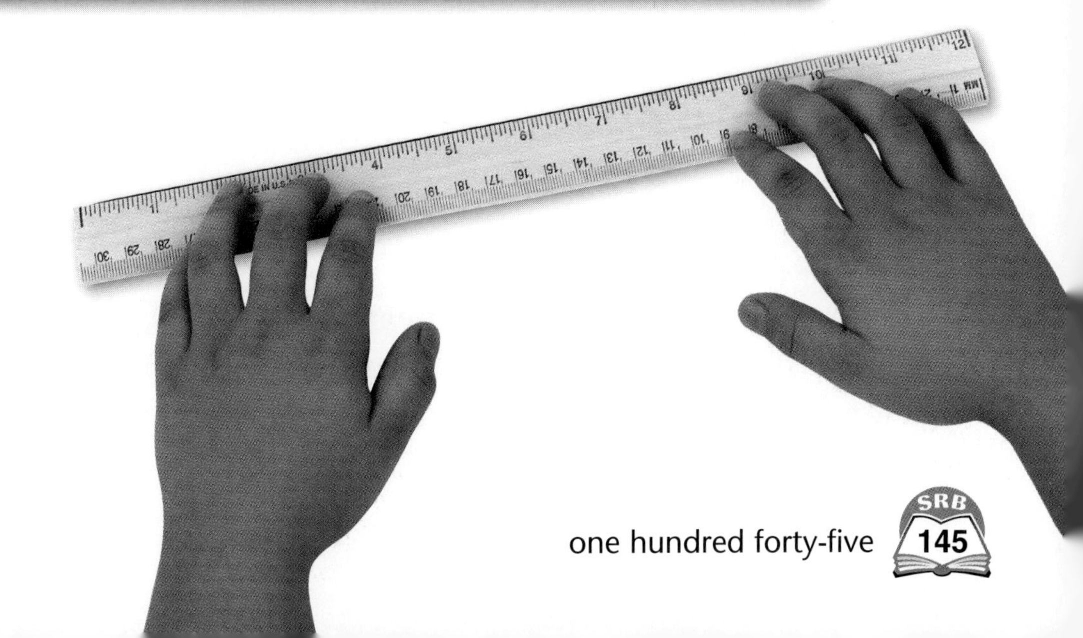

Changing Units of Length in the U.S. Customary System

A basic unit of length in the U.S. customary system is the **inch.** But we measure length using other units, too. **Feet** and **yards** are also used to measure shorter lengths. **Miles** are used to measure longer distances. The table below shows how different units of length compare.

Comparing U.S. Customary Units of Length		Abbreviations for Units of Length
1 foot = 12 inches	1 inch = $\frac{1}{12}$ foot	in. = inch
1 yard = 3 feet	1 foot = $\frac{1}{3}$ yard	ft = foot
1 yard = 36 inches	1 inch = $\frac{1}{36}$ yard	yd = yard
1 mile = 5,280 feet		mi = mile

Did You Know?

The Empire State Building in New York City was built in 1931 and is 1,250 feet tall. The tallest building in the world is Taipei 101 in Taipei, Taiwan. It was built in 2003 and is 1,670 feet tall.

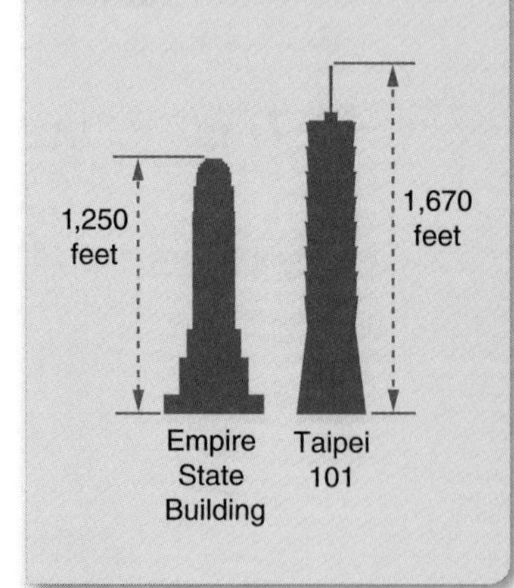

You can use this reference table to change from one unit to another.

Example 6 yards is equal to how many feet?

One yard equals 3 feet.

So 6 yards is equal to 6×3 feet, or 18 feet.

If you know a length in feet, you can change this length to inches. If you know a length in inches, you can change this length to feet. Study the examples on the next page.

Example Ashley is 4 feet 10 inches tall. What is Ashley's height in inches?

One foot equals 12 inches.
So 4 feet is equal to 4×12 inches, or 48 inches.

4 feet + 10 inches is the same as 48 inches + 10 inches, which is 58 inches.

So Ashley is 58 inches tall.

Example Rashid used 36 inches of ribbon to wrap a package. How many feet of ribbon did he use?

One inch is equal to $\frac{1}{12}$ foot.
So 36 inches is equal to $\frac{36}{12}$ feet,

and $\frac{36}{12}$ feet $= 3$ feet.

So Rashid used 3 feet of ribbon.

Example A box is 8 inches long. How many feet long is the box?

One inch is equal to $\frac{1}{12}$ foot.
So 8 inches is equal to $\frac{8}{12}$ foot,

and $\frac{8}{12}$ foot $= \frac{2}{3}$ foot.

So the box is $\frac{2}{3}$ foot long.

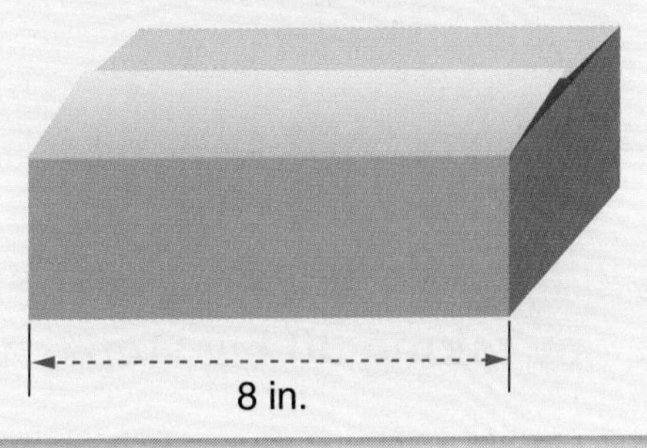

Personal References for U.S. Units of Length

Sometimes you may not have a ruler or yardstick handy. When this happens, you can estimate lengths by using the lengths of common objects and distances that you know. Some examples are given below.

Personal References for U.S. Customary Units of Length

About 1 inch	About 1 foot
Length of a paper clip Width of a quarter Width of a man's thumb	Length of a man's shoe Length of a license plate Length of your math journal
About 1 yard	**About 1 mile**
Width of a door One big step (for an adult) Height of a kitchen counter	2,000 average-size steps (for an adult) Length of 15 football fields (including the end zones)

Example Michael's dad measured the length of a basketball court by taking 30 big steps. About how many feet long is the court?

One big step for an adult is about 1 yard. So the basketball court is about 30 yards long. One yard equals 3 feet.

So the court is about 30×3 feet, or 90 feet long.

Example Alisa's kitchen is as long as 18 math journals. About how many yards long is Alisa's kitchen?

Alisa's math journal is about 1 foot long. So her kitchen is about 18 feet long.

One foot is equal to $\frac{1}{3}$ yard. So 18 feet is equal to $\frac{18}{3}$ yards, and $\frac{18}{3}$ yards = 6 yards.

So Alisa's kitchen is about 6 yards long.

Check Your Understanding

1. **a.** 7 yards = __?__ feet

   b. 2 ft = __?__ in.

   c. 6 feet = __?__ yards

   d. __?__ in. = 2 yd

2. Rashid is 5 feet 3 inches tall. What is Rashid's height in inches?
3. A man's shoe is about as long as __?__ paper clips.
4. 100 big steps for a man is about __?__ yards. About how many feet is that?
5. Michael's uncle walked around a lake. His walk took about 12,000 steps. About how far did he walk?

Check your answers on pages 340 and 341.

yardstick

1-foot ruler

Perimeter

Sometimes we want to know the **distance around** a shape. The distance around a shape is called the **perimeter** of the shape. To measure perimeter, we use units of length such as inches, meters, or miles.

Example Jennifer rode her bicycle once around the edge of a lake.

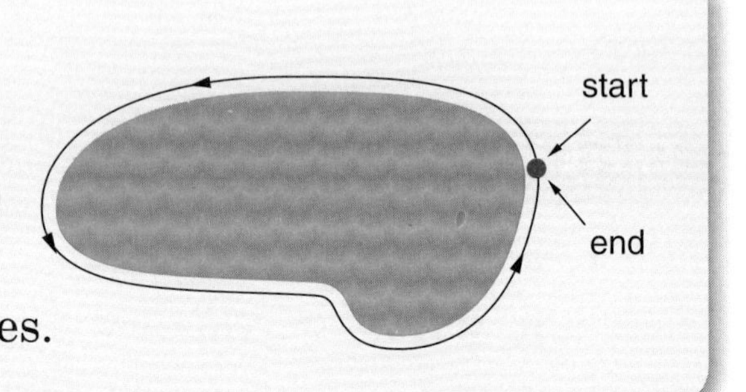

The distance around the lake is 2.3 miles. We say that the perimeter of the lake is 2.3 miles.

To find the perimeter of a polygon, add the lengths of its sides. Always remember to name the unit of length used to measure the shape.

Example Alan ran once around the block. How far did he run?

The distance Alan ran was the perimeter of the block. To find that distance, add the lengths of all four sides.

$$\begin{array}{r} 100 \text{ yd} \\ 60 \text{ yd} \\ 100 \text{ yd} \\ +\ 60 \text{ yd} \\ \hline 320 \text{ yd} \end{array}$$

100 yd

60 yd   60 yd

100 yd

Alan ran 320 yards.

Example Find the perimeter of this square.

All four sides have the same length.

The picture shows that one side is 2 centimeters long.

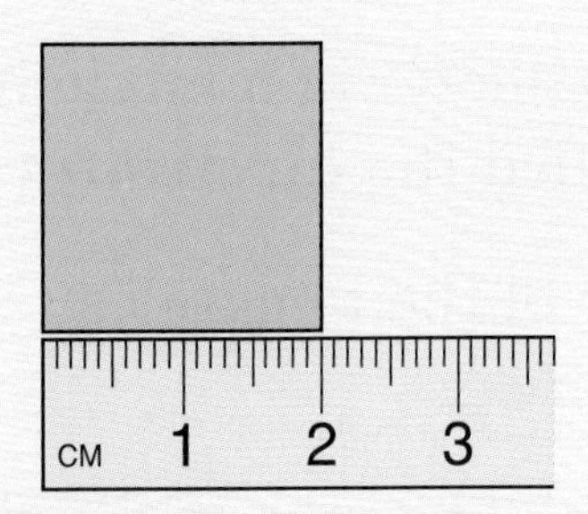

Add the lengths of the four sides.
2 cm + 2 cm + 2 cm + 2 cm = 8 cm

The perimeter of the square is 8 centimeters.

Check Your Understanding

Find the perimeter of the triangle and the square below.

1.

3 ft
5 ft
7 ft

2.

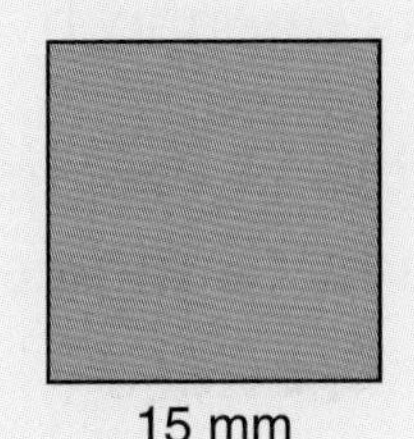

15 mm

3. Measure the sides of your math journal to the nearest half-inch. What is perimeter of your math journal?

Check your answers on page 341.

Circumference and Diameter

The perimeter of a circle is the **distance around** the circle.

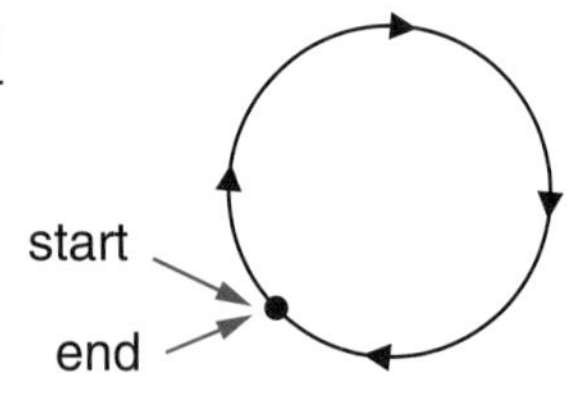

The perimeter of a circle has a special name. It is called the **circumference** of the circle.

The top of the tomato can shown here is shaped like a circle.

The circumference of the circle can be measured with a tape measure.

Wrap the tape measure around the can once. Then read the mark that touches the end of the tape.

The circumference of the can top is how far the can turns when it is opened by a can opener.

The **diameter** of a circle is the distance across the circle, along a segment that goes through the center of the circle.

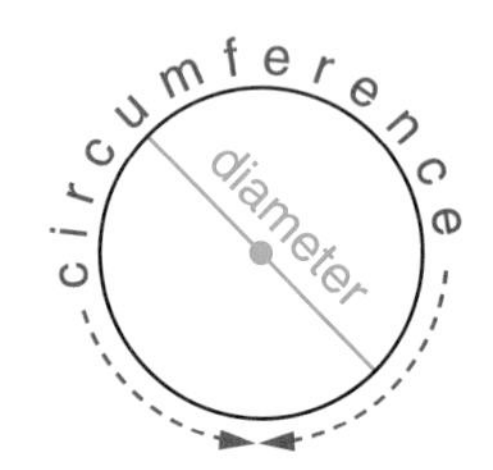

If you know the diameter, there is a simple rule for estimating the circumference.

Circle Rule: The circumference of a circle is just a little more than 3 times the length of the diameter.

Example The diameter of a bicycle wheel is 24 inches. Estimate the circumference of the wheel.

Use the Circle Rule.

The circumference is a little more than 3 × 24 inches, or 72 inches.

Suppose the bicycle tire was cut apart and laid out flat. It would be slightly longer than 72 inches.

Check Your Understanding

1. Measure the diameter of the nickel in millimeters.
2. About how many millimeters is the circumference of the nickel?
3. Which U.S. coin has the smallest circumference?
4. A 12-inch pizza has a diameter of 12 inches. About how many inches is the distance around the pizza?

Check your answers on page 341.

Area

Sometimes we want to know the amount of **surface inside** a shape. The amount of surface inside a shape is called the **area** of the shape.

One way to find the area of a shape is to count the number of squares of a certain size that cover the inside of the shape.

The rectangle below is covered by squares that are 1 centimeter on each side. Each square is called a **square centimeter.**

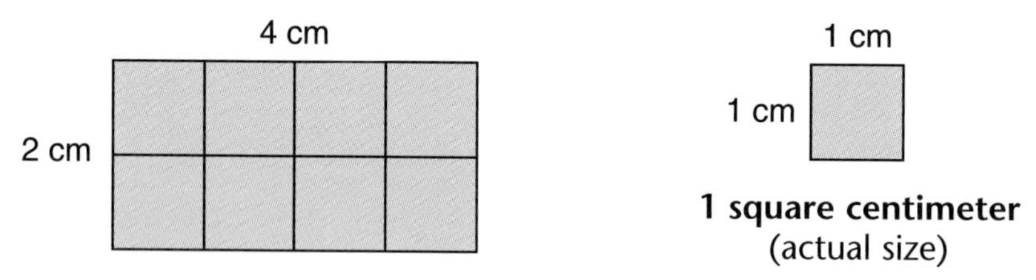

1 square centimeter
(actual size)

Eight of the squares cover the rectangle. The area of the rectangle is 8 square centimeters.

1 in.

1 in.

1 square inch
(actual size)

A square with sides 1 inch long is a **square inch.**

A square with 1-foot sides is a **square foot.**

The **square yard** and **square meter** are larger units of area. They are used to measure large areas such as the area of a floor.

Example Count the square units to find the areas of these shapes.

Each square is 1 square inch.
18 squares cover the rectangle.

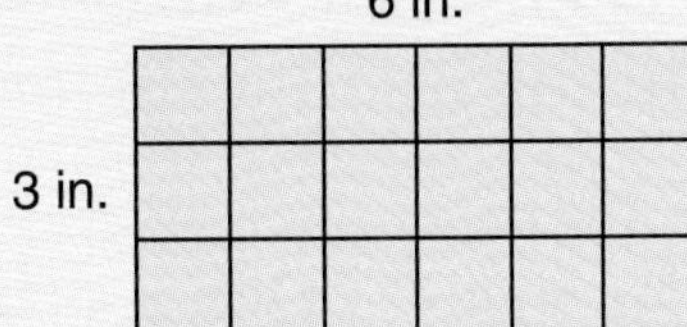

The area of the rectangle is 18 square inches.

Each square is 1 square foot.
14 squares cover the shape.

The area of the shape is 14 square feet.

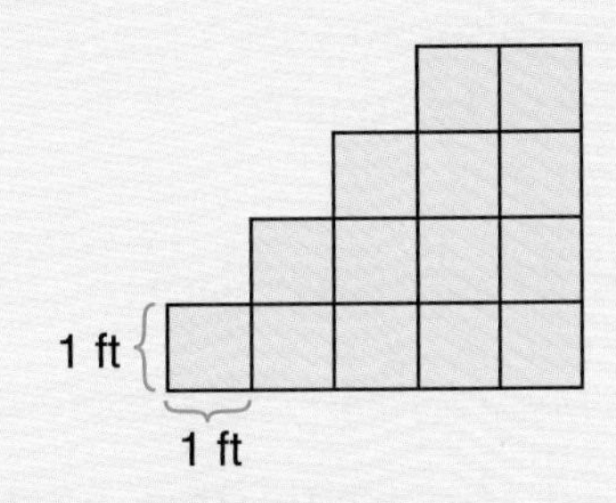

Remember: Perimeter is the **distance around** a shape.
Area is the amount of **surface inside** a shape.

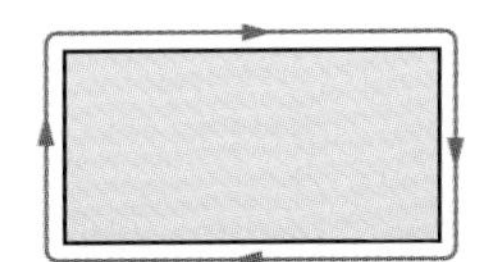

Perimeter is the distance around.

Area is the amount of surface inside.

Some surfaces are too large to cover with squares. It would take too long to count a large number of squares.

To find the area of a rectangle, you do not need to count all of the squares that cover it. The example below shows a shortcut for finding the area.

Example Find the area of this rectangle.

Each square is 1 square foot.

- There are 4 rows of squares.
- Each row has 10 squares.
- So there are 4×10 squares, or 40 squares in all.

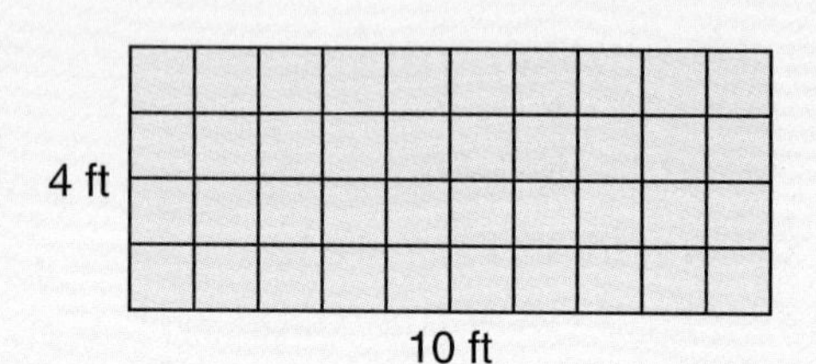

The area is 40 square feet.

Summary To find the area of a rectangle:

1. Count the number of rows.

2. Count the number of squares in 1 row.

3. Multiply: (number of rows) $\times$ (number of squares in 1 row)

Check Your Understanding

Find the area of each rectangle.

1. 2 cm

7 cm

2.

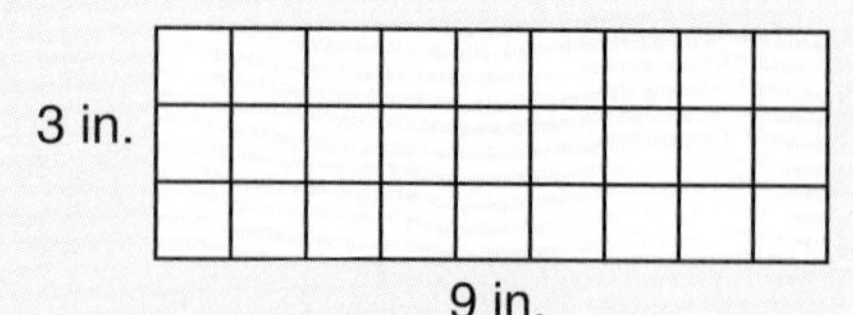

3. Which area is larger, 1 square yard or 1 square meter?

Check your answers on page 341.

Volume

Sometimes we want to know the amount of **space inside** a 3-dimensional object. The amount of space inside an object is called the **volume** of the object. Think of volume as the amount of something an object could hold if the object were hollow on the inside.

We can find the volume of an object by counting the number of cubes of a certain size that would fill the object. To measure the volume of a box, we could fill it with small base-10 cubes.

A base-10 cube has sides that are 1 centimeter long. It is called a **cubic centimeter.** Stack the cubes in the box so that there are no gaps. The volume of the box is the number of cubes needed to fill it.

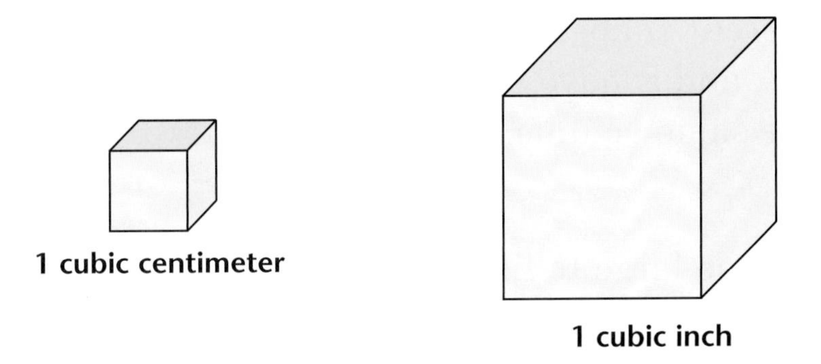

Other cube sizes can be used to measure volume.
A cube with 1-inch sides is called a **cubic inch.**
A cube with 1-foot sides is called a **cubic foot.**
A cube with 1-yard sides is called a **cubic yard.**
A cube with 1-meter sides is called a **cubic meter.**

Example Count cubes to find the volumes of these objects.

Each cube is 1 cubic centimeter.
There are 4 cubes.

The volume is 4 cubic centimeters.

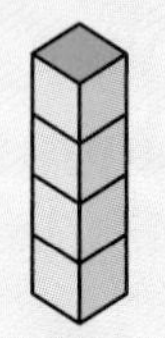

Object A

Each cube is 1 cubic centimeter.
There are 4 cubes.

The volume is 4 cubic centimeters.

Object B

Each cube is 1 cubic foot.
There are 18 cubes.

The volume is 18 cubic feet.

Object C

Objects with different shapes can have the same volume. Objects A and B have different shapes, but they have the same volume.

Sometimes objects are too large to fill with cubes. It would take too long to count a large number of cubes.

A 3-dimensional shape that looks like a box is called a **rectangular prism.** To find the volume of a rectangular prism, you do not need to count all of the cubes that fill it. The example on the next page shows a shortcut method for finding the volume.

Example Find the volume of this rectangular prism.

Each cube is 1 cubic inch.

The top layer has 12 cubes. We cannot see all of the middle layer and the bottom layer. Each of these layers also has 12 cubes.

There are 3 layers of cubes. Each layer has 12 cubes. So there are 3×12 cubes, or 36 cubes in all.

6 in.
2 in.
3 in.

The volume is 36 cubic inches.

Summary To find the volume of a rectangular prism:

1. Count the number of layers.
2. Count the number of cubes in 1 layer.
3. Multiply: (number of layers) $\times$ (number of cubes in 1 layer)

Check Your Understanding

1. Which volume is larger, 1 cubic yard or 1 cubic meter?
2. Find the volume of each stack of cubes.

a.

3 cm
3 cm

b.

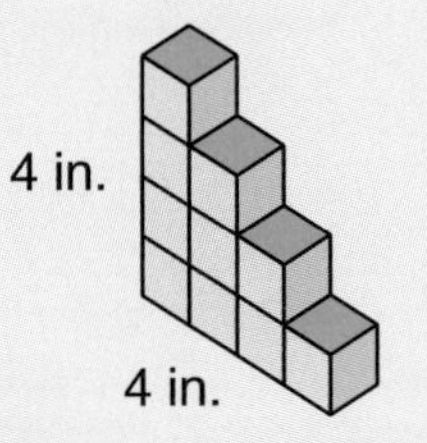

c.

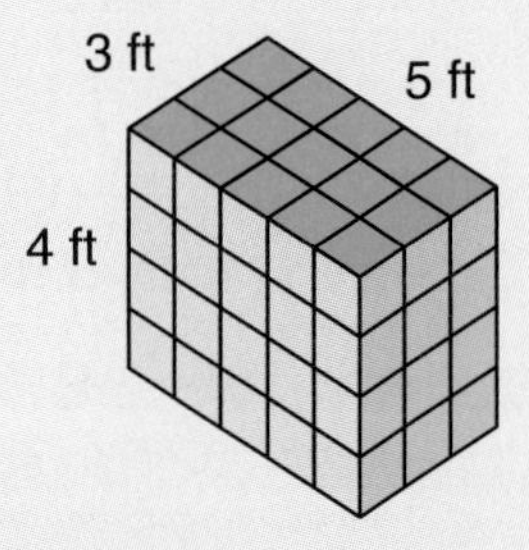

Check your answers on page 341.

Capacity

> **Did You Know?**
> A leaky faucet that drips once a minute can waste 38 gallons of water in a year.

Sometimes we need to know amounts of things that can be poured. All liquids can be poured. Some solids, such as sand and sugar, can be poured, too.

The volume of a container that holds liquids is often called its **capacity.** Capacity is usually measured in units such as **gallons, quarts, pints, cups, fluid ounces, liters,** and **milliliters.**

Liters and milliliters are **metric units.** Gallons, quarts, pints, cups, and fluid ounces are **U.S. customary units.** Most labels for liquid containers give capacity in both metric and U.S. customary units.

The tables below show how different units of capacity compare to each other.

U.S. Customary Units		
1 gallon (gal)	=	4 quarts (qt)
1 gallon	=	2 half-gallons
1 half-gallon	=	2 quarts
1 quart	=	2 pints (pt)
1 pint	=	2 cups (c)
1 cup	=	8 fluid ounces (fl oz)
1 pint	=	16 fluid ounces
1 quart	=	32 fluid ounces
1 half-gallon	=	64 fluid ounces
1 gallon	=	128 fluid ounces

Metric Units		
1 liter (L)	=	1,000 milliliters (mL)
1 milliliter	=	$\frac{1}{1,000}$ liter

cup

pint

quart

half-gallon

gallon

You can use the tables to change from one unit to another.

Example Change 3 quarts to pints, cups, and gallons.

1 quart equals 2 pints. So 3 quarts is 3×2 pints, or 6 pints.

1 quart equals 2 pints; each pint equals 2 cups.
So 1 quart equals 4 cups, and 3 quarts is 3×4 cups, or 12 cups.

4 quarts equals 1 gallon. So 1 quart equals $\frac{1}{4}$ gallon,
and 3 quarts is $3 \times \frac{1}{4}$ gallon, or $\frac{3}{4}$ gallon.

$3 \text{ qt} = 6 \text{ pt} = 12 \text{ c} = \frac{3}{4} \text{ gal}$

One liter is a little more than 1 quart.

Example About how many cups is 2 liters?

1 liter is about 1 quart. 2 liters is about 2 quarts.
Since 1 quart equals 4 cups, 2 quarts is 2×4 cups, or 8 cups.

So 2 liters is about 8 cups.

Example A 2-liter bottle of water contains about how many fluid ounces?

2 liters is about 8 cups (by the example above).
1 cup equals 8 fluid ounces.
So 8 cups is 8×8 fluid ounces, or 64 fluid ounces.

A 2-liter bottle contains about 64 fluid ounces.

Check Your Understanding

1. **a.** 10 pt = __?__ c **b.** 8 pt = __?__ qt **c.** 12 qt = __?__ gal
2. 10 liters is about how many cups?

Check your answers on page 341.

Weight

Did You Know?

The first steam-powered train was built in 1804. It pulled a load of 24,000 pounds (12 tons) of iron.

Today, in the United States, we use two different sets of standard units to measure weight.

- The standard unit for weight in the metric system is the **gram.** A small, plastic base-10 cube weighs about 1 gram. Heavier weights are measured in **kilograms.** One kilogram equals 1,000 grams.
- Two standard units for weight in the U.S. customary system are the **ounce** and the **pound.** Small weights are measured in ounces, and heavier weights are measured in pounds. One pound equals 16 ounces. Some weights are reported in both pounds and ounces. For example, we might say that "the box weighs 4 pounds 3 ounces."

The tables below list the units of weight most often used. They show how these units compare to each other.

Metric Units	
1 gram (g) = 1,000 milligrams (mg)	1 metric ton (t) = 1,000 kilograms
1 milligram = $\frac{1}{1,000}$ gram	1 kilogram = $\frac{1}{1,000}$ metric ton
1 kilogram (kg) = 1,000 grams	
1 gram = $\frac{1}{1,000}$ kilogram	

U.S. Customary Units	
1 pound (lb) = 16 ounces (oz)	1 ton (T) = 2,000 pounds
1 ounce = $\frac{1}{16}$ pound	1 pound = $\frac{1}{2,000}$ ton

Example Arthur's bowling ball weighs 10 pounds 5 ounces. How many ounces does the ball weigh?

One pound equals 16 ounces. So 10 pounds is 10×16 ounces, or 160 ounces. 10 pounds 5 ounces is 160 ounces + 5 ounces, or 165 ounces.

So the bowling ball weighs 165 ounces.

How can we compare two weights, such as 6 ounces and 280 grams, that use different units? One weight uses U.S. customary units, and the other weight uses metric units.

Use the items shown below to compare 1 gram and 1 ounce and to compare 1 pound and 1 kilogram.

small, plastic base-10 cube	30 base-10 cubes	1 pint of strawberries	$2\frac{1}{4}$ pints of strawberries

about 1 gram	about 1 ounce	about 1 pound	about 1 kilogram

Example A volleyball weighs 280 grams. A softball weighs 6 ounces. Which ball weighs more?

One ounce equals about 30 grams. So 6 ounces is about 6×30 grams, or 180 grams. The softball weighs about 180 grams.

So the volleyball weighs more than the softball.

Check Your Understanding

Is 600 grams heavier than 1 pound?

Check your answer on page 341.

The following pages show samples of different kinds of scales. The **capacity** and the **precision** are given for each scale.

The **capacity** of a scale is the greatest weight that the scale can hold. For example, most infant scales have a capacity of about 25 pounds. An infant scale would not be used to weigh a third grader.

The **precision** of a scale is the accuracy of the scale. If you can read a weight on an infant scale to the nearest ounce, then the precision for that scale is 1 ounce. With a balance scale, you can measure and compare weights to the nearest gram. A balance scale is much more precise than an infant scale because a gram is much lighter than an ounce.

Some scales are extremely precise. They can weigh things that cannot even be seen with the naked eye. Other scales are very large. They can be used to weigh objects that weigh as much as 1,000 tons (2,000,000 pounds). Many scales display weights in both metric and U.S. customary units.

Check Your Understanding

1. Arrange these weights from lightest to heaviest:

1 pound 1 gram 1 kilogram 1 ounce

2. Copy and complete.

a. __?__ lb = 16 oz

b. 1,000 mg = __?__ g

c. 1 kg = __?__ g

d. 600 g = __?__ kg

e. 10 lb = __?__ oz

f. 8,000 lb = __?__ T

3. What is the accuracy of a scale called?

Check your answers on page 341.

Samples of Scales

Types of scales will vary in capacity and precision.

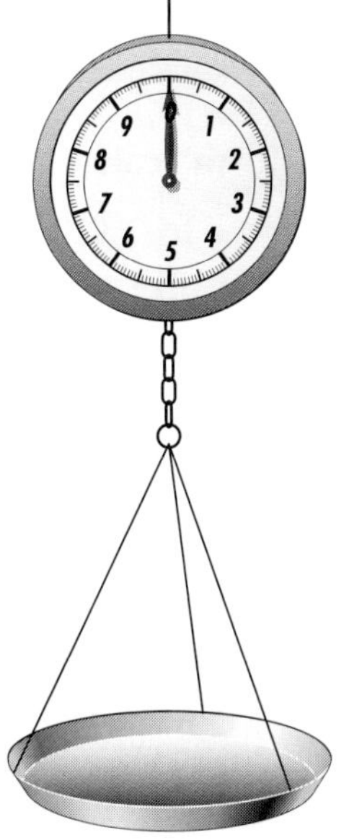

produce scale
capacity: 10 lb
precision: 1 oz

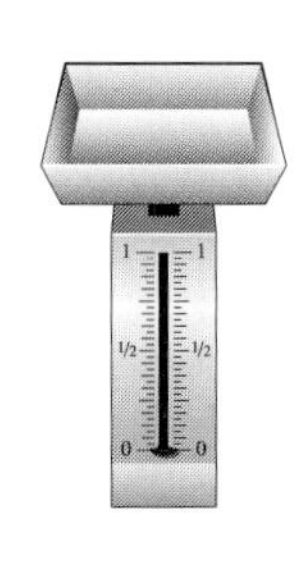

capacity: 16 oz
precision: $\frac{1}{2}$ oz

food scales

capacity: 12 lb
precision: 1 oz

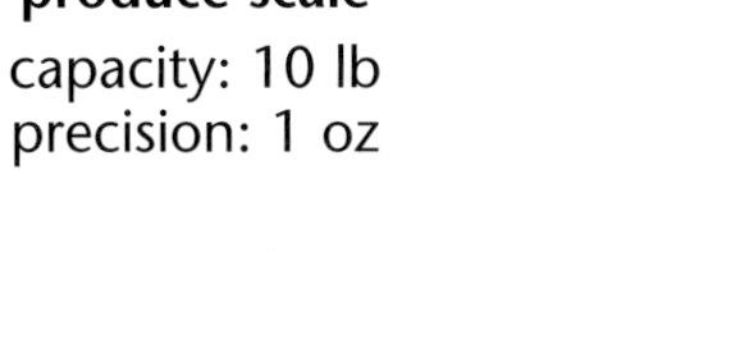

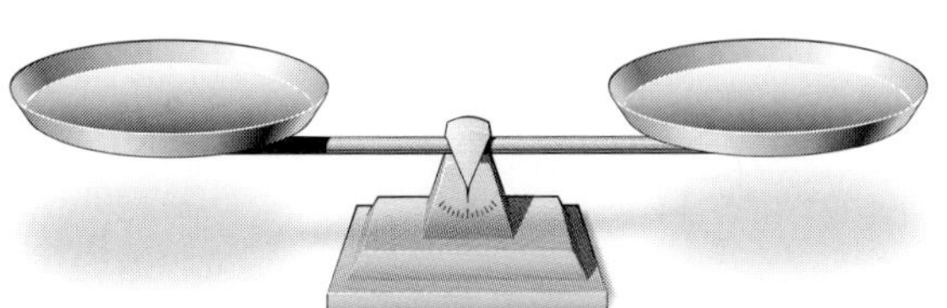

balance scale
capacity: 2 kg
precision: 1g

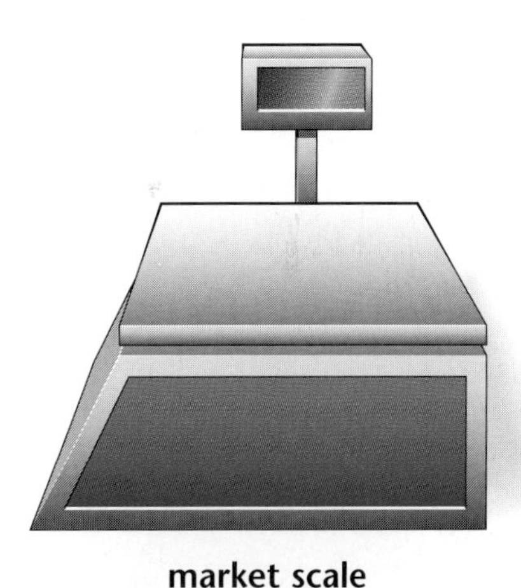

market scale
capacity: 30 lb or 15 kg
precision: 0.01 lb or 5 g

weight set for balance scale

package scale
capacity: 70 lb
precision: 1 lb

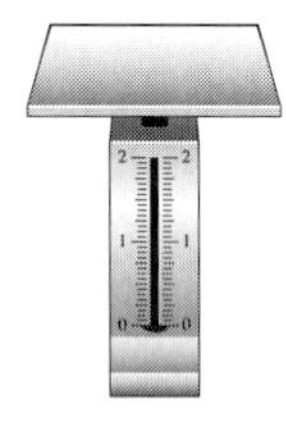

letter scale
capacity: 2 lb
precision: 1 oz

platform scale
capacity: 1 T to 1,000 T
precision: $\frac{1}{4}$ lb to 1 T

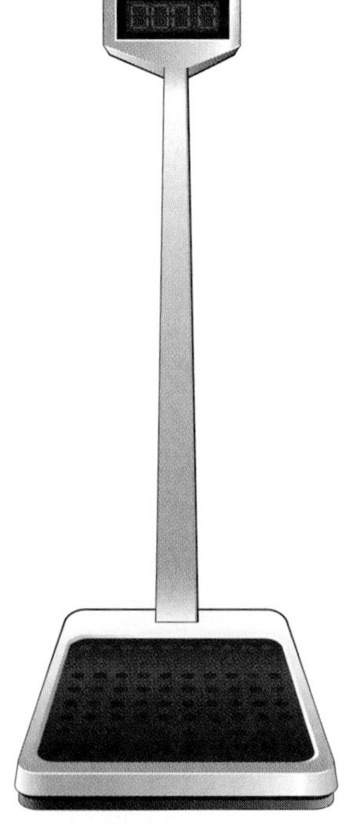

doctor's scale
capacity: 250 lb or 115 kg
precision: 0.8 oz or 25 g

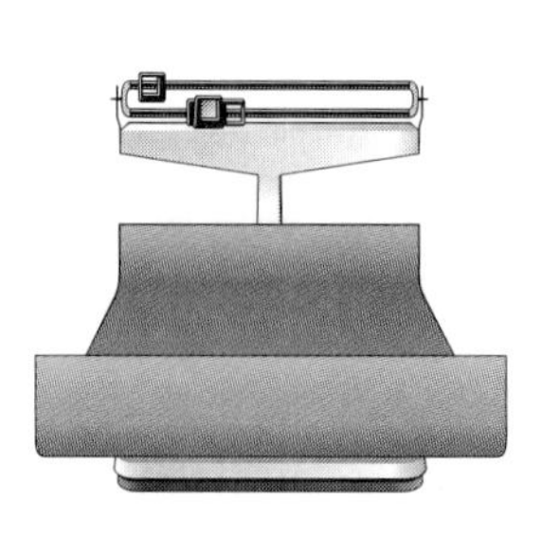

infant scale
capacity: 25 lb
precision: 1 oz

bath scale
capacity: 300 lb or 135 kg
precision: 0.1 lb or 50 g

spring scales

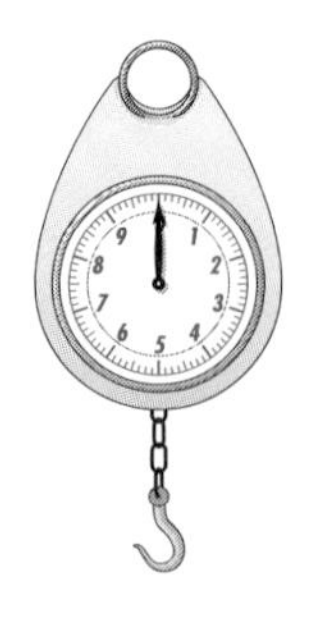

capacity: 10 oz
precision: $\frac{1}{8}$ oz

capacity: 500 g
precision: 20 g

Measuring Angles

The Babylonians lived about 3,000 years ago in what is now the country of Iraq. The way we measure angles was invented by the Babylonians. They counted a year as having 360 days. They used this same number, 360, for measuring angles. The angle measurer shown below is their invention.

The circle is divided into 360 equal parts called **degrees.** The numbers printed in the circle are written with a small raised circle (°). The small circle is a symbol for the word *degree.* For example, we read 270° as "270 degrees."

Think of walking around the circle. Start at the 0° mark. Walk clockwise around the circle. The degree numbers written in the circle show how many marks on the circle you have traveled past.

Example When you walk from 0° to 45°, you pass 45 marks along the circle.

When you walk from 0° to 180°, you pass 180 marks. You pass half of the 360 marks on the circle. You are halfway around the circle.

When you walk from 0° to 360°, you pass all 360 marks. You are back where you started.

Here is how to use an angle measurer (protractor) to measure angles.

1. Place the hole in the center of the measurer over the vertex of the angle. The vertex is the point where the sides of the angle meet.
2. Line up the 0° mark on the measurer with the side of the angle where the small, curved arrow begins.
3. Find where the other side of the angle crosses the measurer. Read the degree measure.

Example The small, curved arrow shows which angle is being measured.

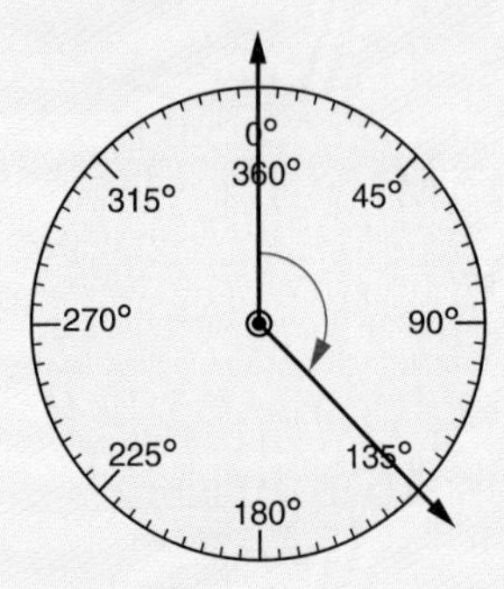

The angle measures 135°.

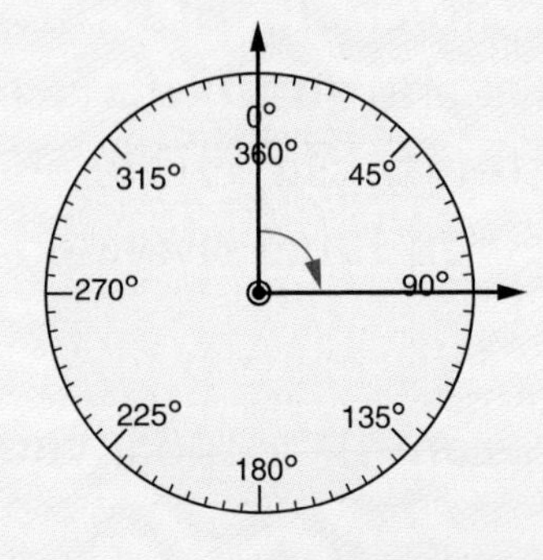

The angle measures 90°.

A 90° angle is $\frac{1}{4}$ turn on the circle.

A 90° angle is called a **right angle.**

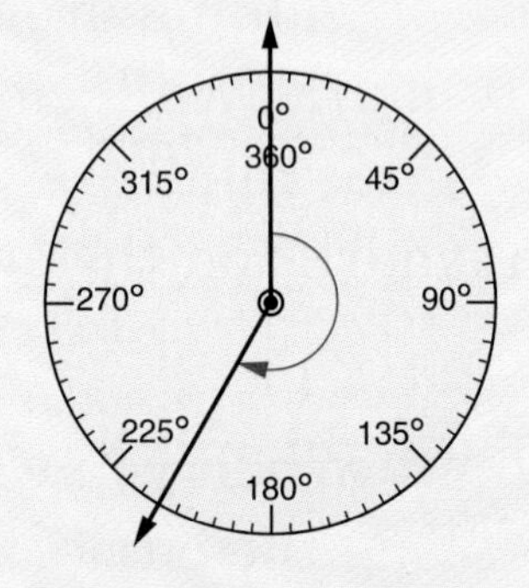

The angle measures between 180° and 225°.

The angle measure is closer to 225° than it is to 180°.

Reference Frames

Temperature

The **temperature** of something is how hot or cold it is. A **thermometer** measures temperature. The common thermometer is a glass tube that contains a liquid. When the temperature goes up, the liquid expands and moves up the tube. When the temperature goes down, the liquid shrinks and moves down the tube.

In the U.S. customary system, temperature is measured in **degrees Fahrenheit (°F).** In the metric system, temperature is measured in **degrees Celsius (°C).**

Did You Know?

Lightning can heat the air to temperatures of 50,000°F and above.

Example Water freezes at 32°F or 0°C. Water boils at 212°F or 100°C.

A small thermometer for taking body temperatures usually has marks that are spaced $\frac{2}{10}$ (0.2) of a degree apart. This allows you to make very accurate measurements of body temperature.

92 93 94 95 96 97 98 99 100 101 102 103 104 105 106 107 108

98.6°

Normal body temperature is about 98.6°F.

Temperatures may be negative numbers. The temperature −12°F is read as "negative 12 degrees," or as "12 degrees below 0."

Most thermometers have marks that are spaced 2 degrees apart.

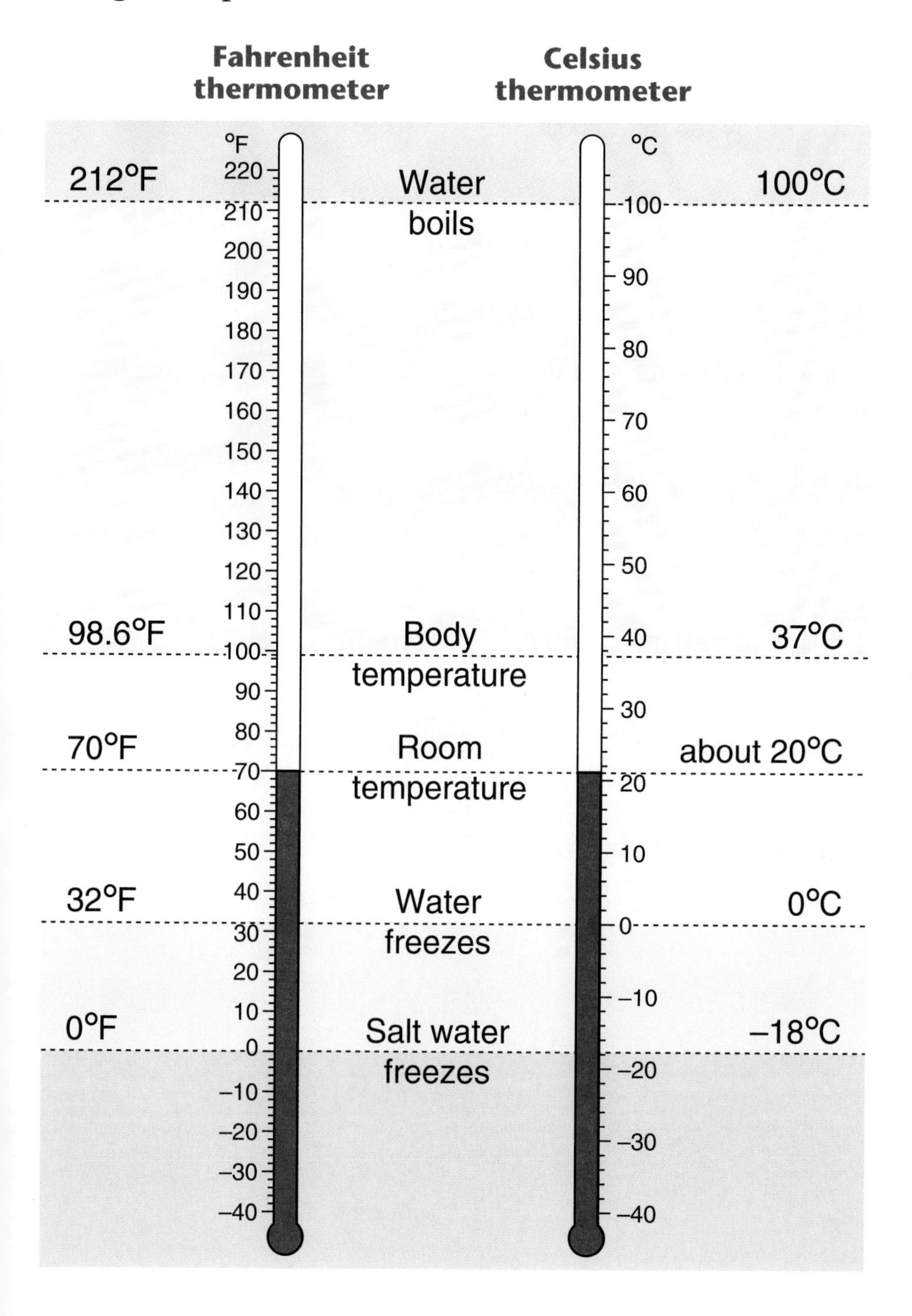

Sometimes you want to find the difference between two temperatures.

Example Find the temperature difference between 48°F and 94°F.

One way to find the difference is to start with the smaller number and count up to the larger number. Start with 48. Add 2 to get 50. Then add 40 to get 90. Then add 4 to get 94. The total added is $2 + 40 + 4$, or 46.

Another way to find the difference is to subtract.
$94 - 48 = 46$

The difference is 46 degrees, or 46°F.

Example Find the temperature difference between 52°C and −20°C.

Start with the negative temperature, −20. Add 20 to get 0. Then add 52 to get 52. The total added is $20 + 52$, or 72.

The difference is 72 degrees, or 72°C.

Sometimes the temperature may change, and you want to find the new temperature.

Example The temperature was 50°F at 6:00 P.M. By 9:00 P.M. the temperature had gone down 20 degrees. What was the temperature at 9:00 P.M.?

Since the temperature went down, subtract 20 from the starting temperature.
$50 - 20 = 30$

The temperature was 30°F at 9:00 P.M.

Check Your Understanding

1. Find the missing temperatures.

	Fahrenheit degrees (°F)	Celsius degrees (°C)
a. Water boils	212°F	? °C
b. Water freezes	? °F	0°C
c. Body temperature	? °F	37°C
d. Room temperature	70°F	? °C
e. Salt water freezes	0°F	? °C

2. Find the temperature difference between 22°F and 60°F.

3. Find the temperature difference between −30°C and 50°C.

4. The temperature was 73°F at noon. By 4:00 P.M. the temperature had fallen 35°F. What was the temperature at 4:00 P.M.?

Check your answers on page 342.

Time

We use **time** in two ways:

1. to tell when something happens, and

2. to tell how long something takes or lasts.

Example Marta goes to sleep at 9:30 P.M. She wakes up at 7:15 A.M. Marta has slept for 9 hours and 45 minutes.

9:30 P.M. and 7:15 A.M. are times that tell when something happens.
9 hours and 45 minutes tells how long Marta's sleep lasted.

A.M. is an abbreviation that means "before noon." It refers to the period from midnight to noon. **P.M.** is an abbreviation that means "after noon." It refers to the period from noon to midnight. Noon is written as 12:00 P.M. Midnight is written as 12:00 A.M.

The table shows how units of time compare.

Units of Time

1 minute = 60 seconds	1 year = 52 weeks plus 1 day, or 52 weeks plus 2 days in leap years
1 hour = 60 minutes	1 year = 365 days, or 366 days in leap years
1 day = 24 hours	1 decade = 10 years
1 week = 7 days	1 century = 100 years
1 month = 28, 29, 30, or 31 days	1 millennium = 1,000 years
1 year = 12 months	

Did You Know?

A Galapagos tortoise has a life span of 200+ years.

The life span of insects usually falls between 2 weeks and 8 months.

One second is a very short time. But we often want to measure times to a fraction of a second. Some wristwatches have a stopwatch that records times to the nearest $\frac{1}{100}$ (or 0.01) second.

Example Running events in the Olympics are timed to the nearest 0.01 second. In the 2004 Summer Olympics, Justin Gatlin won the 100-meter run. His time was 9.85 seconds.

Check Your Understanding

1. What do A.M. and P.M. stand for?
2. Put these times in order, starting with midnight:

   7:30 P.M.   12:00 A.M.   12:00 P.M.   4:15 A.M.

   9:45 P.M.   10:50 A.M.   3:05 P.M.   2:55 A.M.
3. How many years are in 1 decade?
4. How many years are in 3 centuries?

Copy and complete.

5. 1 day = _?_ hours
6. 4 weeks = _?_ days
7. **Challenge:** How many seconds are in 1 hour?

Check your answers on page 342.

Calendars

Earth revolves around the sun. It takes 365 days, 5 hours, 48 minutes, and 46 seconds to make one complete revolution. This time is the exact meaning of one year.

We use a **calendar** to keep track of the days of each week and month in a year. The calendars for most years show 365 days. But every four years, we add an extra day in February. These special years are called **leap years.** Each leap year has 366 days.

Years that are leap years follow a pattern. The pattern has these two rules:

1. Any year that can be divided by 4 (with no remainder) is a leap year. So 2004, 2008, and 2012 are all leap years.

2. Years that end in 00 are special cases. They are leap years only when they can be divided by 400 (with no remainder). So 2000 and 2400 are leap years. But 1900 and 2100 are *not* leap years.

Number of Days in Each Month

January	31 days	August	31 days
February	28 or 29* days	September	30 days
March	31 days	October	31 days
April	30 days	November	30 days
May	31 days	December	31 days
June	30 days		
July	31 days	*29 days in leap year	

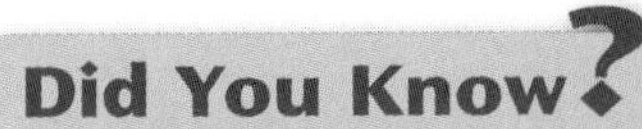

It takes 247.7 years for Pluto to make 1 complete revolution around the sun.

Here is a calendar for the year 2008. This is a leap year.

2008

JANUARY

S	M	T	W	T	F	S
		1	2	3	4	5
6	7	8	9	10	11	12
13	14	15	16	17	18	19
20	21	22	23	24	25	26
27	28	29	30	31		

FEBRUARY

S	M	T	W	T	F	S
					1	2
3	4	5	6	7	8	9
10	11	12	13	14	15	16
17	18	19	20	21	22	23
24	25	26	27	28	29	

MARCH

S	M	T	W	T	F	S
						1
2	3	4	5	6	7	8
9	10	11	12	13	14	15
16	17	18	19	20	21	22
23	24	25	26	27	28	29
30	31					

APRIL

S	M	T	W	T	F	S
		1	2	3	4	5
6	7	8	9	10	11	12
13	14	15	16	17	18	19
20	21	22	23	24	25	26
27	28	29	30			

MAY

S	M	T	W	T	F	S
				1	2	3
4	5	6	7	8	9	10
11	12	13	14	15	16	17
18	19	20	21	22	23	24
25	26	27	28	29	30	31

JUNE

S	M	T	W	T	F	S
1	2	3	4	5	6	7
8	9	10	11	12	13	14
15	16	17	18	19	20	21
22	23	24	25	26	27	28
29	30					

JULY

S	M	T	W	T	F	S
		1	2	3	4	5
6	7	8	9	10	11	12
13	14	15	16	17	18	19
20	21	22	23	24	25	26
27	28	29	30	31		

AUGUST

S	M	T	W	T	F	S
					1	2
3	4	5	6	7	8	9
10	11	12	13	14	15	16
17	18	19	20	21	22	23
24	25	26	27	28	29	30
31						

SEPTEMBER

S	M	T	W	T	F	S
	1	2	3	4	5	6
7	8	9	10	11	12	13
14	15	16	17	18	19	20
21	22	23	24	25	26	27
28	29	30				

OCTOBER

S	M	T	W	T	F	S
			1	2	3	4
5	6	7	8	9	10	11
12	13	14	15	16	17	18
19	20	21	22	23	24	25
26	27	28	29	30	31	

NOVEMBER

S	M	T	W	T	F	S
						1
2	3	4	5	6	7	8
9	10	11	12	13	14	15
16	17	18	19	20	21	22
23	24	25	26	27	28	29
30						

DECEMBER

S	M	T	W	T	F	S
	1	2	3	4	5	6
7	8	9	10	11	12	13
14	15	16	17	18	19	20
21	22	23	24	25	26	27
28	29	30	31			

Example Thanksgiving is always on the fourth Thursday in November. So Thanksgiving Day is November 27 in 2008.

January 1, 2008 is a Tuesday. So the last day of 2007 (December 31, 2007) is a Monday.

Check Your Understanding

1. Which months have 31 days?
2. What day of the week is January 1, 2009?
3. What date does the fourth Monday of May, 2008, fall on?
4. What is the day of the week and the date one week after May 30, 2008?

Check your answers on page 342.

Seasons and Length of Daytime

The year is divided into four seasons.

Dates	Season North of Equator	Season South of Equator
Dec 22 to Mar 20	Winter	Summer
Mar 21 to Jun 20	Spring	Fall
Jun 21 to Sep 21	Summer	Winter
Sep 22 to Dec 21	Fall	Spring

Daytime is the part of a day from sunrise to sunset. **Nighttime** is the part of a day from sunset to sunrise. The length of one day is always 24 hours. But the length of daytime is not always the same.

- In most places on Earth, the length of daytime changes from one day to the next day.
- At any place on the equator, daytime is always 12 hours long and nighttime is always 12 hours long.

On March 21 and September 22, daytime and nighttime have the same length. They are each 12 hours long everywhere on Earth.

The first day of winter has the shortest daytime and longest nighttime of the year. This is December 22 in the United States. The first day of winter is June 21 for any place south of the equator. The first day of summer has the longest daytime and shortest nighttime of the year. This is June 21 in the United States, and December 22 for any place south of the equator.

The length of daytime depends upon the time of year (the date). It also depends upon how far you are from the equator. The table below shows the length of daytime at some different places and times of year.

Length of Daytime (in hours and minutes)

Date	Equator	Houston, Texas	Seward, Alaska	North Pole
March 21	12 hr 0 min	12 hr 0 min	12 hr 0 min	12 hr 0 min
June 21	12 hr 0 min	14 hr 4 min	18 hr 49 min	24 hr 0 min
September 22	12 hr 0 min	12 hr 0 min	12 hr 0 min	12 hr 0 min
December 22	12 hr 0 min	10 hr 14 min	5 hr 54 min	0 hr 0 min

Example Compare Houston and Seward on June 21. Both cities are north of the equator, so June 21 is the first day of summer in both places.

- For Houston, June 21 has the longest daytime of the year (14 hr 4 min).
- For Seward, June 21 has the longest daytime of the year (18 hr 49 min).
- Seward is much farther from the equator than Houston is. And Seward gets about 5 more hours of sunlight than Houston does on June 21.

Check Your Understanding

1. What are the beginning and ending dates for spring in the place where you live?
2. Compare Houston and Seward on December 22.

Check your answers on page 342.

Coordinate Grids

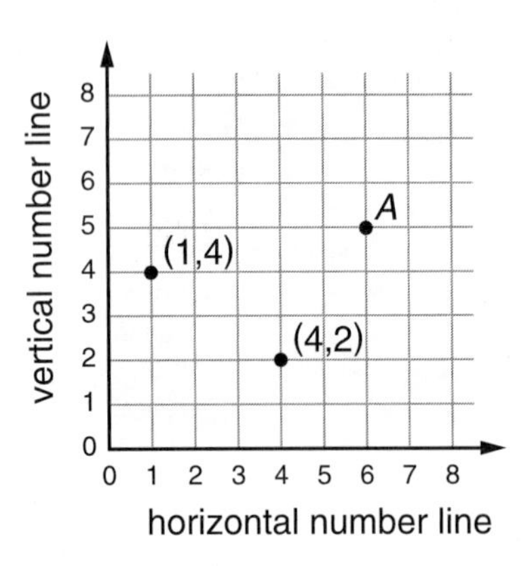

Sometimes we use two number lines to make a **coordinate grid.**

We locate points on the coordinate grid with two numbers. The numbers are written in parentheses.

Example Locate the point (4,2) on the coordinate grid.

Start at the point marked 0 on the horizontal number line. Move along the horizontal number line to the point marked 4. Then move up from there to the line that is named 2 on the vertical number line.

The point where you stop is named (4,2).

Example Name the location of point *A*.

Find the number on the horizontal number line that is directly under point *A*. That number is 6. Then find the number on the vertical number line that is directly to the left of point *A*. That number is 5.

The name for point *A* is (6,5).

Pairs of numbers like (6,5) and (1,4) are called **ordered pairs.**

The numbers in parentheses are called the **coordinates** of the point. The numbers 6 and 5 are the coordinates of the point (6,5).

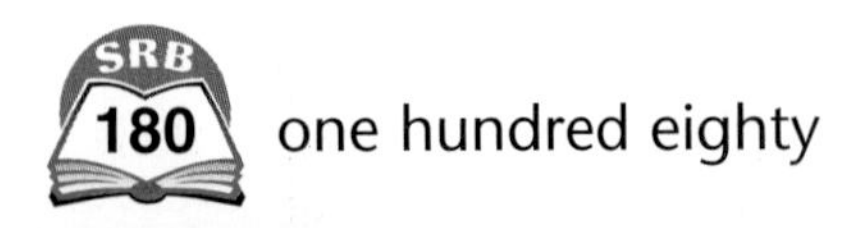

Check Your Understanding

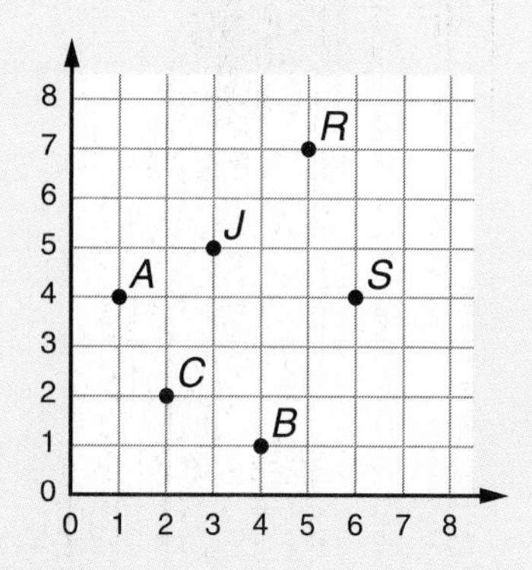

1. Which letter names the point at each location?

   a. (4,1) **b.** (3,5)

2. Write the location of each point.

   a. point C **b.** point R

3. Use the coordinate grid on the campground map to write the coordinates for each place.

   a. Campground (1,2)
   b. Fire Lookout (___,___)
   c. Lodge (___,___)
   d. Playground (___,___)
   e. Ski Jump (___,$5\frac{1}{2}$)
   f. Swimming Pool (___,___)

Campground Map

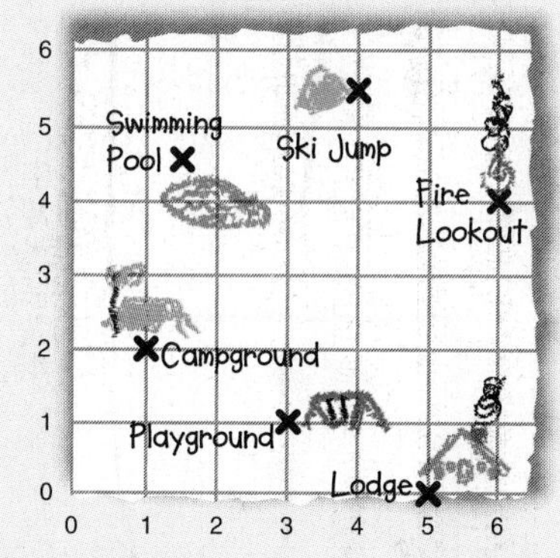

Check your answers on page 342.

Scale Drawings

Maps and drawings often have a **distance key.** The distance key tells how to change the distances on a map or drawing to real-life distances. Sometimes the distance key is called a **scale.**

Example Josh used grid paper to draw a map. The distance key on Josh's map shows that every 1 inch on his map equals 100 feet of real distance.

Josh walks from his house to school by going along Main Street and then along Ellis Avenue. On his map it is 1 inch from his house to Ellis Avenue. It is another 3 inches along Ellis Avenue to the school. The total distance from Josh's house to school is 4 inches. Each inch equals 100 feet, so the school is 400 feet from Josh's house.

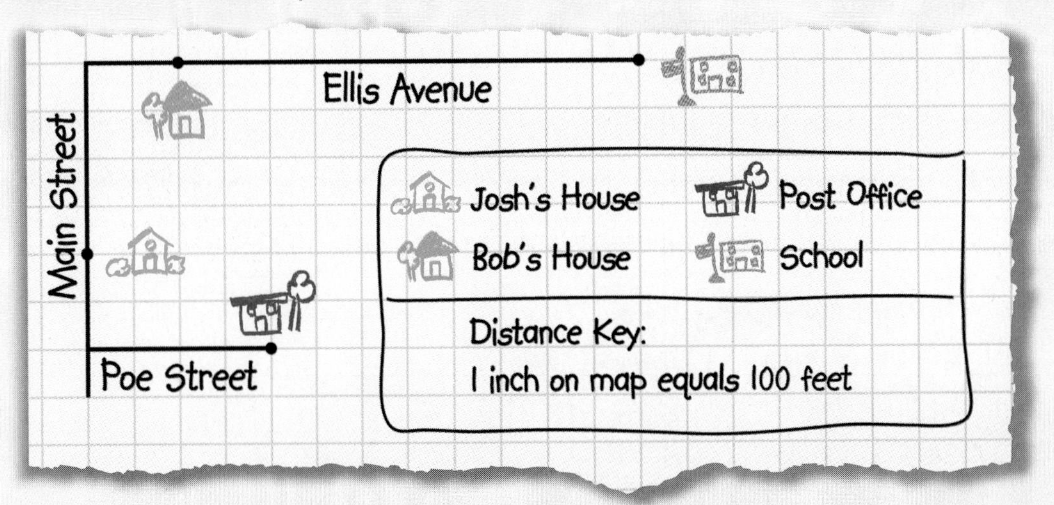

The grid paper Josh used has 4 grid squares per inch. So every grid square is $\frac{1}{4}$-inch long and stands for 25 feet of real distance.

Bob's house is 6 grid square sides away from Josh's house. So Bob's house is 6×25 feet, or 150 feet away from Josh's house.

Timekeeping

Mathematics ... Every Day

In the past, people gauged the passage of time by observing nature. They kept track of the seasons by paying attention to migrating birds, falling leaves, and changes in sun, moon, and star positions.

Some birds migrate back to their breeding grounds when the weather begins to get warmer.

The leaves of many types of trees change color when cooler weather arrives.

Constellations, like Orion, appear in different parts of the sky during different seasons.

Tools for Tracking Time

At some point, people figured out how to make tools to track the passage of time. Early tools relied on the movements of celestial bodies like the sun, moon, and stars.

These huge stones in England, known as Stonehenge, were set up nearly 5,000 years ago. Stonehenge can measure when summer and winter begin.

As early as 3,500 B.C., the Egyptians and Sumerians built shadow clocks. As the sun moves across the sky, the length and direction of the shadow give a rough idea of the time.

Advancements in shadow clocks led to the sundial. Sundials can be very accurate, but they have limitations. For example, they do not work on very cloudy days.

Water Clocks

In ancient times, people also used timepieces that did not rely on the movement of celestial bodies. One of these timepieces is known as the water clock.

Simple water clocks fed water at a constant speed into a container. It was possible to read the time by measuring the level of water in the container.

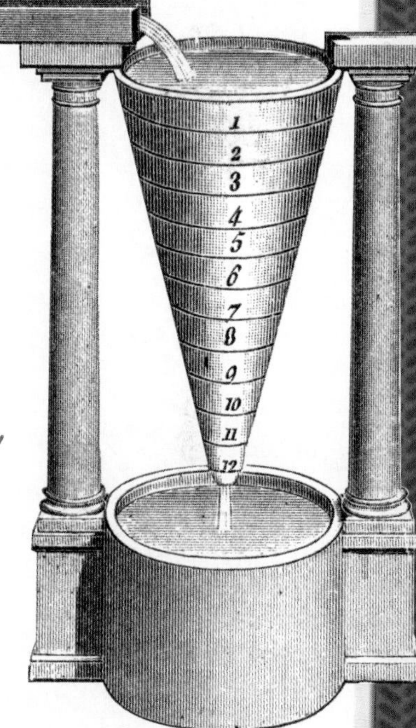

The clepsydra is a mechanical water clock. It was developed around 500 B.C. It is designed so that water runs constantly out of one vessel and into another. Unlike sundials, clepsydras work in cloudy weather and in the dark.

This is a model of the Su Sung Water Clock, built in China in 1088. At thirty feet tall, it was probably the largest and most elaborate water clock ever built.

Pendulums and Pendulum Clocks

An important advance in clock-making came about with the invention of the pendulum clock.

In the 1580s, the Italian scientist Galileo Galilei made an important discovery. He noticed that it took the same amount of time for a chandelier to complete one swing no matter how wide the swing.

About 75 years later, Dutch scientist Christiaan Huygens used Galileo's discovery to make the first pendulum clock. These clocks used a swinging pendulum to keep the time.

The pendulum clock shown here was actually built a little later, in the 1750s.

Mathematics ... Every Day

Modern Clocks

Clocks have become more accurate and more versatile over time.

In the 1920s, timekeeping took a giant leap forward as inventors took advantage of a property of quartz crystals. Running electricity through these crystals causes them to vibrate at a constant rate, like the constant rate of a pendulum swing.

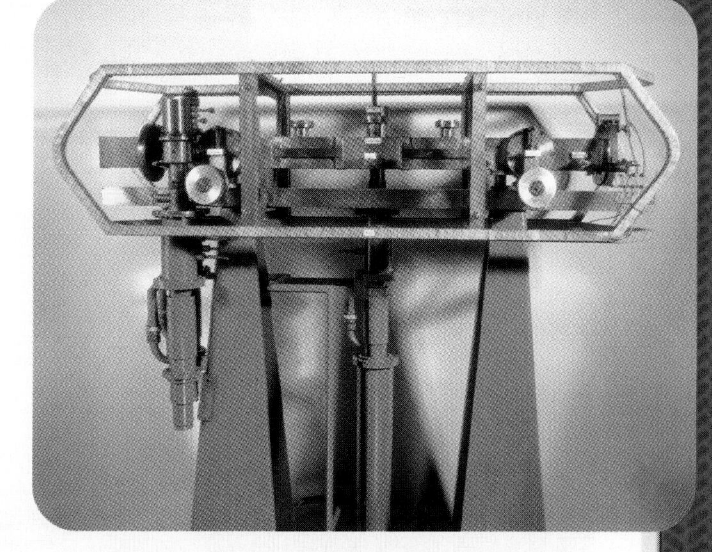

Today, quartz clocks are built into many watches, calculators, and personal computers. They are, by far, the most popular timekeepers because they are inexpensive and accurate.

In 1955, scientists built a clock that is more accurate than a quartz clock. It is called the cesium atomic clock. It is accurate to 1 second every 1,000,000 years!

Back to Nature

Although we rely on technology for accurate timekeeping, we still use nature to estimate the time.

By gazing at our own shadows, we can tell whether it is morning, noon, or evening. In the evening and morning, the sun is closer to the horizon, so our shadows are long. Around noon, the sun is higher in the sky, so our shadows are shorter. ➤

➤ We use many clues to figure out what time of day it is. What time of day do you think it is in this photo?

In what ways do you track the passage of time?

Estimation

When You Have to Estimate

An **estimate** is an answer that should be close to an exact answer. You make estimates every day.

- You estimate how long it will take to drive from one place to another.
- You estimate how much money you will need to buy some things at the store.
- You estimate how many inches you will grow in the next year.

It may be impossible to find an exact answer. When this happens, you *must* estimate the answer.

Did You Know?

Other people besides weather forecasters use estimates to make predictions. Sportscasters predict the scores of future games. News writers and reporters predict election results.

Example Weather forecasters predict temperatures for the next day. They must estimate because they don't know what the exact temperatures will be.

They use words like **expect, predict,** and **about.** These words let people know that they are giving estimates, not exact amounts.

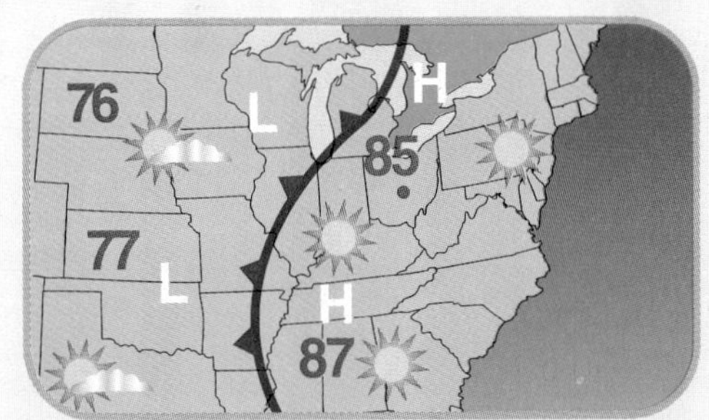

"Columbus, Ohio may *expect* sunny weather tomorrow. A high temperature of *about* 85 degrees is *predicted*."

Some estimates are called **ballpark estimates.** A ballpark estimate is an answer that may not be close to the exact answer, but is close enough to help you solve a problem.

Estimate When You Don't Need an Exact Answer

An estimate may help you answer a question, so that you do not need to find an exact answer.

Example Carlie has $5.00. Is that enough money to buy a $1.39 bottle of juice and a $2.89 salad?

Carlie can estimate. She can use simple numbers that are close to the exact prices.

	Exact prices	**Simple numbers that are close**
$2.89 is almost $3.	$2.89	$3.00
$1.39 is almost $1.50.	$1.39	+$1.50
$3 + $1.50 equals $4.50.		$4.50

Carlie has enough money to buy the juice and the salad.

Example Ming read 13 pages in half an hour. About how long will it take him to read 38 pages?

Estimate how long it will take Ming. Use simple numbers that are close to the exact numbers.

	Exact numbers	**Simple numbers that are close**
13 is close to 10.	13 pages	10 pages
38 is close to 40.	38 pages	40 pages

Reading 40 pages should take about 4 times as long as reading 10 pages.

It will take Ming about 2 hours to read 38 pages.

Estimate to Check Calculations

Sometimes you do want an exact answer. Making an estimate can help you check your answer. Your estimate should be close to the exact answer. If your estimate is not close, you know that you should try the calculation again.

Example Tanesha took a trip. On Monday she went 316 miles. On Tuesday she went 447 miles. On Wednesday she went 489 miles. Tanesha added the three numbers and got 975.

Tanesha made a ballpark estimate to check her answer. She used simple numbers that were close to the numbers in the problem.

	Exact numbers	Simple numbers that are close
316 is close to 300.	316	300
447 is close to 400.	447	400
489 is close to 500.	489	+ 500
Add the 3 simple numbers.		1,200

Tanesha knows that her answer of 975 must be wrong. She added the three numbers again. This time she got 1,252.

The new answer makes more sense. It is close to Tanesha's estimate of 1,200.

Adjusting Numbers

Try to use simple numbers when you estimate. The simple numbers should be close to the exact numbers in the problem.

Here is one way to **adjust** a number and get a simpler number.

1. Keep the first digit of the number.

2. Replace the other digits of the number by zeros.

Examples

Exact number	Adjusted number
124	100
77	70
5,100	5,000
138,429	100,000
2	2

Examples Each problem below is written again using adjusted numbers. The adjusted numbers are used to find an estimate. The estimates are circled.

Exact numbers	Adjusted numbers	Exact numbers	Adjusted numbers
347	300	452	400
− 212	− 200	+ 86	+ 80
	(100)		(480)

A more accurate way to adjust numbers is to round the numbers. Here is a way to **round a number.**

Examples

		Round 68	Round 529
Step 1	Write the number you are rounding.	68	529
Step 2	Keep the first digit. Replace the other digits by zeros. This is the **lower number.**	60	500
Step 3	Add 1 to the first digit. This is the **higher number.**	70	600
Step 4	Is the number you are rounding closer to the lower number or the higher number?	higher	lower
Step 5	Round to the closer of the two numbers.	70	500

Sometimes the number you are rounding is halfway between the lower number and the higher number. When this happens, round to the higher number.

Example You are rounding 45.

The lower number is 40. The higher number is 50. 45 is halfway between 40 and 50.

So round 45 to 50, which is the higher number.

Check Your Understanding

Round each number.

1. 78 **2.** 34 **3.** 85 **4.** 555 **5.** 4,302

6. Estimate the sum of 282 + 47 by using rounded numbers.

Check your answers on page 342.

Patterns and Functions

Picture Patterns

Shapes are often arranged in regular ways to form patterns. Floor and ceiling tiles often form a pattern.

The pictures below show some different ways that bricks are used to build walls. Each way of laying bricks forms a different pattern.

Sometimes a picture pattern is given. You are asked to continue the pattern. You must decide what the next picture in the pattern will be.

To find the next picture for a pattern, you will have to guess. Your guess should be a good guess, not a wild guess. Look carefully at what is given and try to find a pattern. Use the pattern to help guess what the next picture will be.

Did You Know?

A kaleidoscope is a tube that contains mirrors and pieces of colored glass. As you look into it and rotate the tube, you see changing patterns that are always symmetrical.

Example What is the next picture in this pattern?

?

The pattern shows a series of squares. Each square contains one dot. The dot moves clockwise from one corner to the next.

So the next picture will look like this: 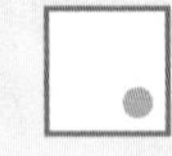

Here are some other picture patterns. There is enough information in each pattern to make a good guess about the next picture.

Example Find the next picture in this pattern.

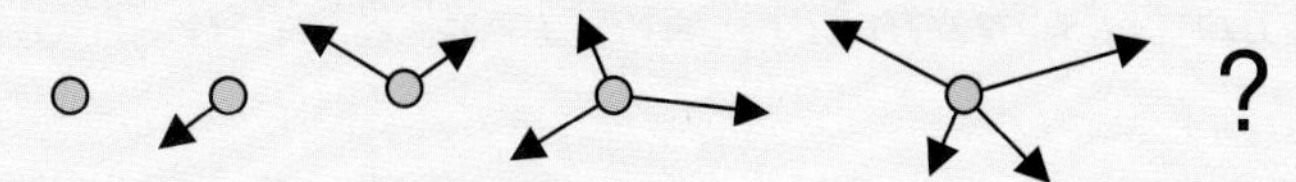

Each picture in the series has a small circle.
The number of arrows increases by one each time.

So the next picture will have 5 arrows. 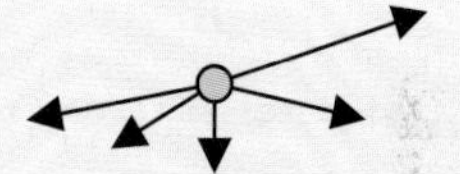

Example Find the next picture in this pattern. 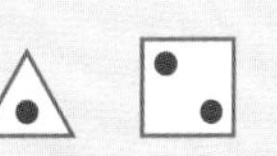 

Each picture has two parts. One part is a polygon. Each polygon in the pattern has one more side than the polygon before it. The last polygon is a pentagon (5 sides). So the next polygon will be a hexagon (6 sides).

The other part of each picture is a set of dots inside the polygon. Each polygon has one more dot than the polygon before it. The pentagon has 3 dots. So the next polygon will have 4 dots.

The next picture for this pattern is 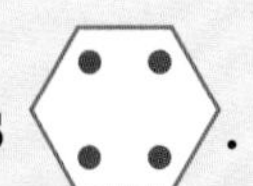.

Check Your Understanding

Draw the next picture in each pattern.

1. 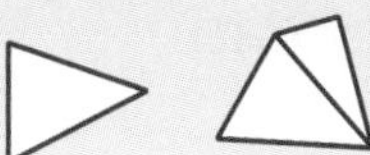 

2. 

Check your answers on page 343.

Number Patterns

Dot pictures can be used to represent numbers. The dot pictures can help us find number patterns.

All of the dot pictures shown here are for counting numbers. The *counting numbers* are 1, 2, 3, and so on.

Even numbers

2 4 6 8 10 12

Even numbers are counting numbers that have a remainder of 0 when they are divided by 2. An even number has a dot picture with 2 equal rows.

Odd numbers

1 3 5 7 9 11 13

Odd numbers are counting numbers that have a remainder of 1 when they are divided by 2. An odd number has a dot picture with 2 equal rows, plus 1 extra dot.

Triangular numbers

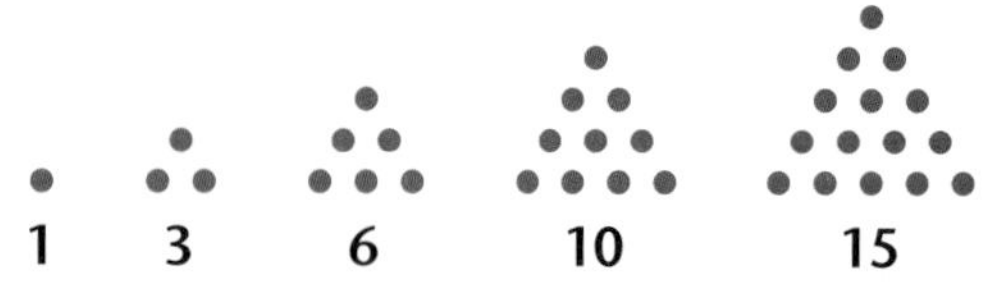

Each dot picture has a triangular shape, with the same number of dots on each side. Each row has 1 more dot than the row above it. Any counting number that has a dot picture like one of these is called a **triangular number.**

Square numbers

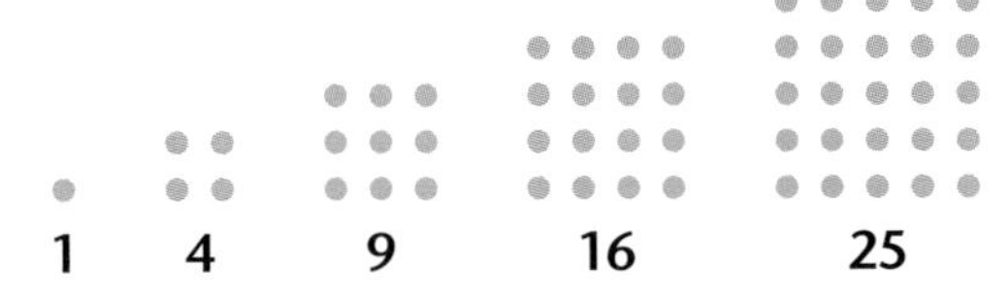

A **square number** is the product of a counting number multiplied by itself. For example, 16 is a square number because 16 equals $4 * 4$ or 4^2. A square number has a dot picture with a square shape, with the same number of dots in each row and column.

Rectangular numbers

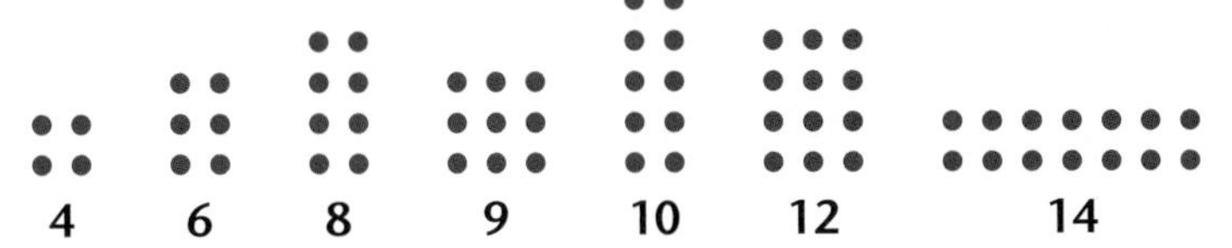

A **rectangular number** is a counting number that is the product of 2 smaller counting numbers. For example, 12 is a rectangular number because $12 = 3 * 4$. A rectangular number has a dot picture with a rectangular shape, with at least 2 rows and at least 2 columns.

Prime numbers

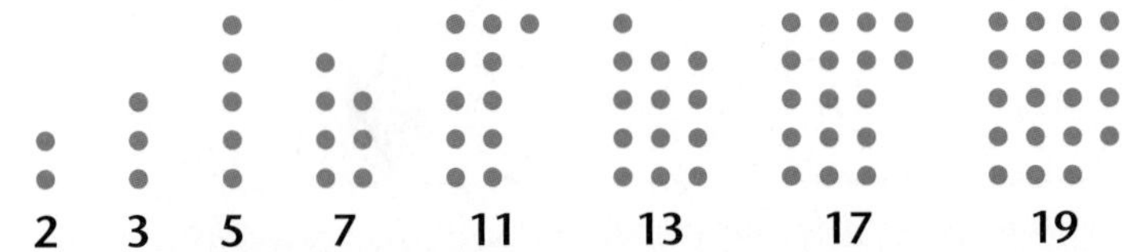

A **prime number** is a counting number greater than 1 that is *not* equal to the product of 2 smaller counting numbers. So a prime number cannot be a rectangular number. This means that a prime number cannot be fit into a rectangular shape (with at least 2 rows and at least 2 columns).

Frames and Arrows

A Frames-and-Arrows diagram is one way to show a number pattern. This type of diagram has three parts:

- a set of **frames** that contain numbers;
- **arrows** that show the path from one frame to the next frame;
- a box with an arrow below it. The box has a **rule** written inside. The rule tells how to change the number in one frame to get the number in the next frame.

Example Here is a Frames-and-Arrows diagram.

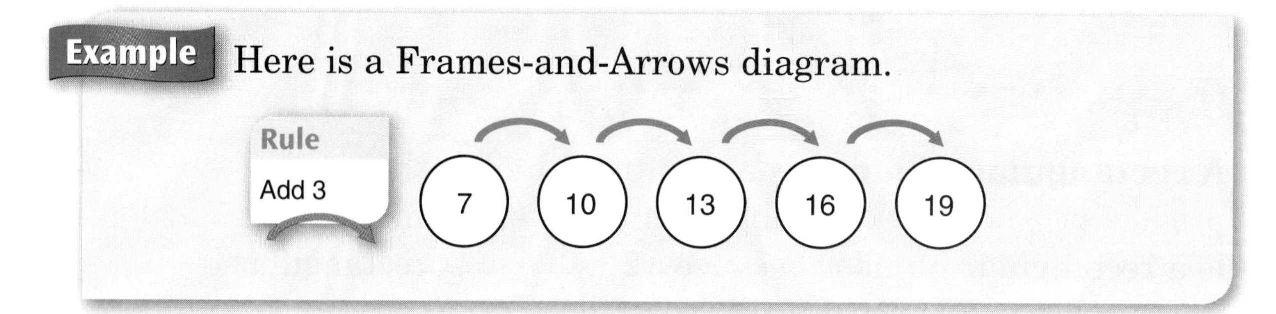

Example Use the rule to fill in the empty frames.

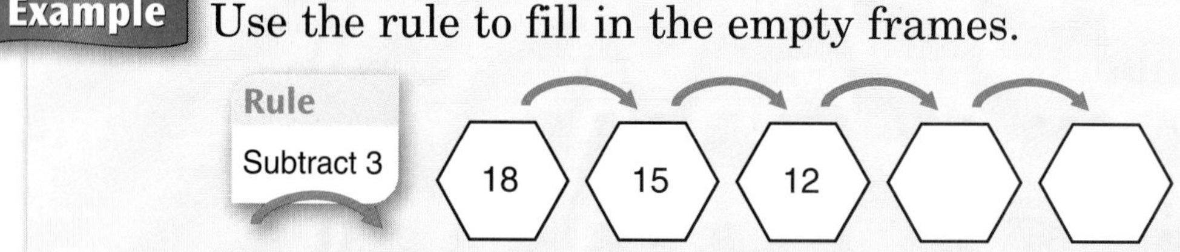

The rule is "Subtract 3." Look at the frame with the number 12. If you subtract 3 from 12, the result is 9. Write 9 in the next frame. Then subtract 3 from 9. The result is 6. Write 6 in the last frame. The filled-in diagram looks like this.

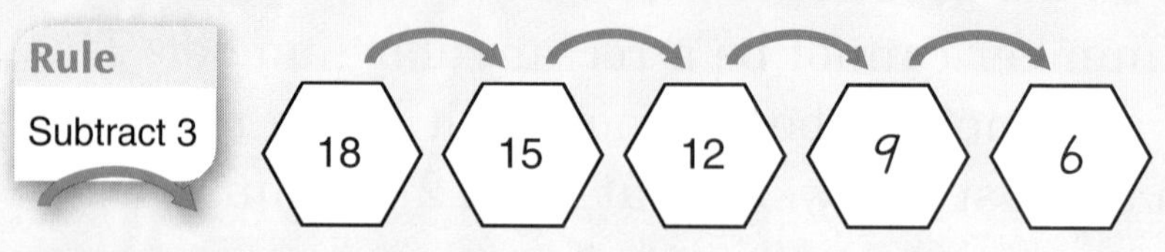

Sometimes the rule is not given. You must use the numbers in the frames to find the rule.

Example Find the rule for this diagram.

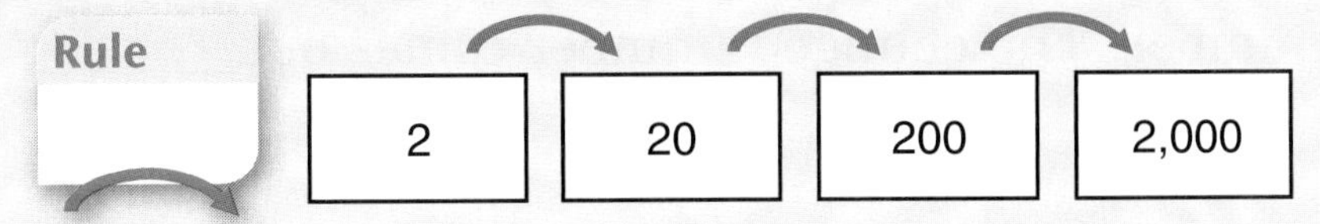

Each number is 10 times larger than the number in the frame that comes before it.

So the rule is "Multiply by 10," or "× 10."

Sometimes the rule is not given and the frames are not all filled in. Find the rule first. Then use the rule to fill in the empty frames.

Example Find the rule and fill in the empty frames.

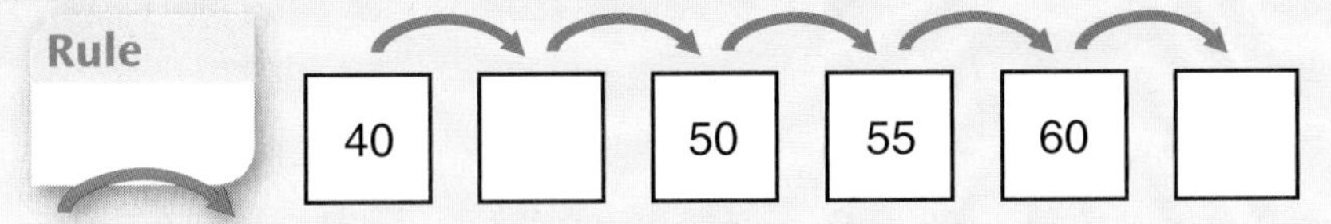

The numbers 50, 55, and 60 can help you to find the rule. Each number is 5 more than the number in the frame before it.

So the rule is "Add 5."

Now use the rule to fill in the empty frames. The second frame will contain the number 40 + 5, or 45. The last frame will contain the number 60 + 5, or 65. The filled-in diagram looks like this.

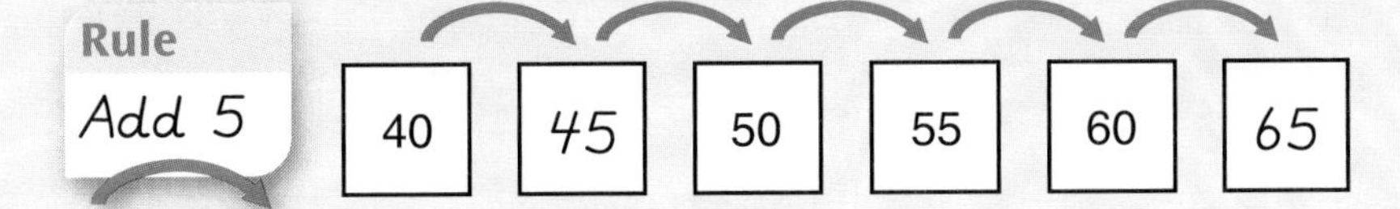

Function Machines

A **function machine** is an imaginary machine. The machine is given a rule for changing numbers. You drop a number into the machine. The machine uses the rule to change the number. The changed number comes out of the machine.

Here is a picture of a function machine.
The machine has been given the rule "+4."
The machine will add 4 to any number that is put into it.

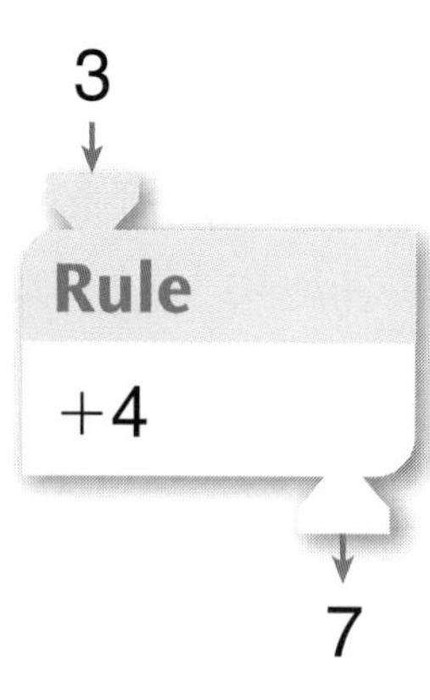

Example If you drop 3 into the function machine above, it will add 3 + 4. The number 7 will come out.

If you drop 1 into the machine, it will add 1 + 4. The number 5 will come out.

If you drop 0 into the machine, it will add 0 + 4. The number 4 will come out.

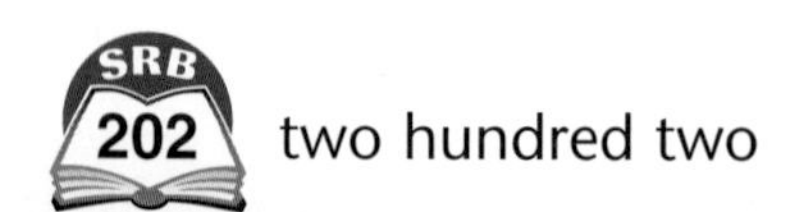

Function Machines and "What's My Rule?"

You can use a table of **in** and **out** numbers to keep track of the way a function machine changes numbers.

Write the numbers that are put into the machine in the **in** column.

in	out
0	4
1	5
2	6
3	7

Write the numbers that come out of the machine in the **out** column.

Example The rule is +10. You know the numbers that are put into the machine. Find the numbers that come out of the machine.

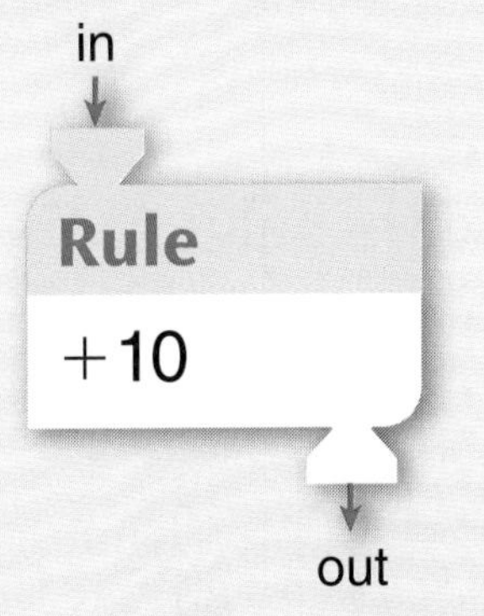

in	out
27	
61	
148	

If 27 is put in, then 37 comes out.
If 61 is put in, then 71 comes out.
If 148 is put in, then 158 comes out.

Example The rule is -7. You know the numbers that come out of the machine. Find the numbers that were put into the machine.

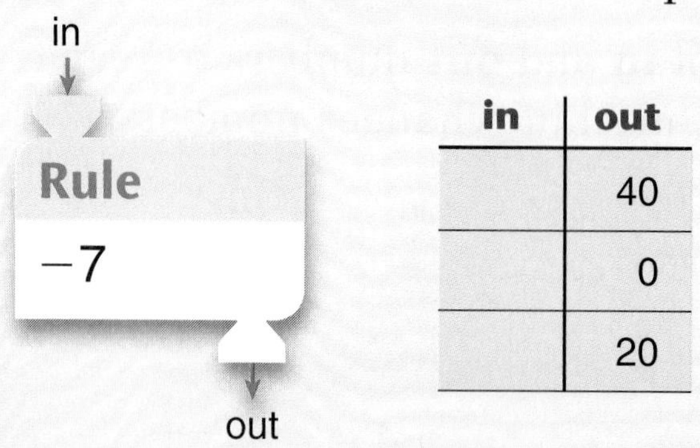

in	out
	40
	0
	20

The machine subtracts 7 from any number that is put into it.
The number that comes out is always 7 less than the number put in.

If 40 comes out, then 47 was the number put in.
If 0 comes out, then 7 was the number put in.
If 20 comes out, then 27 was the number put in.

If you have a table of some **in** and **out** numbers, you can find the rule.

Example The rule is not known. Use the table to find the rule.

in	out
30	35
75	80
101	106

Each number in the **out** column is 5 more than the number in the **in** column.

The rule is Add 5 or $+5$.

Patterns in the Wild

Scientists study patterns that occur in the world around us, including patterns in the lives of wild animals.

Migration Patterns

Migration patterns are cycles of movement that many animals in the wild repeat each year.

Gray whales complete one of the longest migration routes of any mammal, averaging 10,000–14,000 miles. ➤

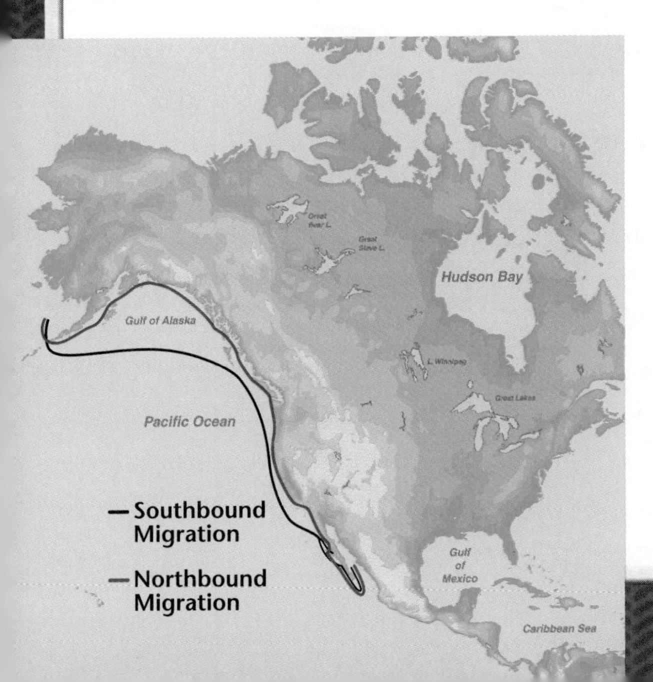

➤ Each fall, the whales leave the cold waters of the north where they feed. They travel south for 2–3 months to the Sea of Cortez, where they breed and give birth. They return north in the spring.

Patterns of Growth

From birth to adulthood the sizes of animals increase in fairly predictable ways. Some animals, like many fish and reptiles, have growth rates that are like a multiplying number pattern (1, 2, 4, 8, 16...). They start out tiny and mature to be thousands of times their birth size.

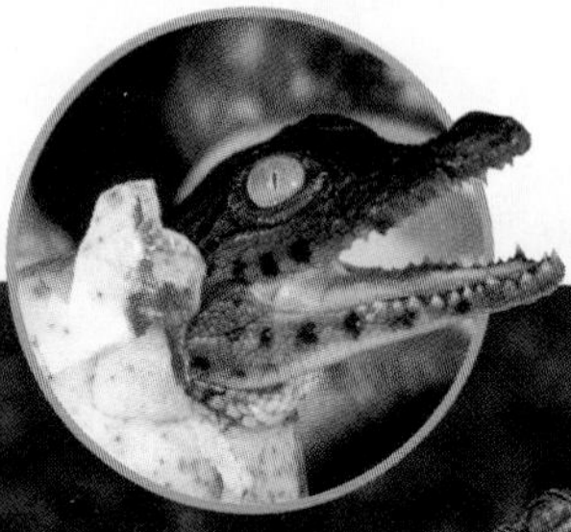

A crocodile hatchling weighs about 2 ounces.

A mature female crocodile can easily weigh 1,000 pounds. This is 8,000 times heavier than the hatchling.

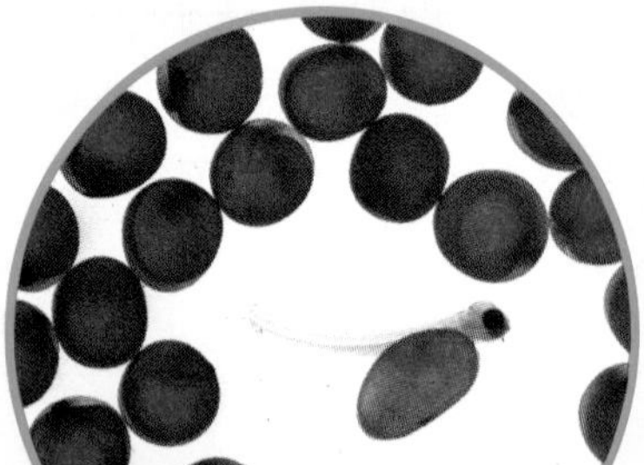

A salmon fry can weigh a mere $\frac{1}{100}$ of an ounce.

A full-grown salmon can weigh over 50 pounds. This is 90,000 times heavier than a fry.

Other animals, like many mammals, have growth rates that are like an adding number pattern (1, 3, 5, 7, 9....). They start out larger and don't grow nearly as fast.

A newborn human infant weighs about 8 pounds. The baby's mother may weigh 140 pounds—only about 18 times more than her baby.

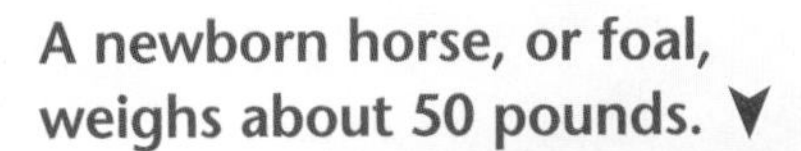

A newborn horse, or foal, weighs about 50 pounds.

The foal's mother weighs about 1,500 pounds—only 30 times more than the foal.

Patterns of Survival

All animals strive to reach adulthood, or maturity. The number of babies an animal has is related to the number that is likely to survive. Some animals, such as many mammals and birds, have a pretty good chance of making it.

Mammals often give birth to one to ten live young. Female mammals, like this lioness, will fiercely protect their offspring. With her protection, more than half of these cubs will live to reach maturity.

Ducks lay about 5 to 15 eggs and keep them warm until they hatch.

This duck does her best to protect her ducklings. If all goes well, about 1 out of 4 of them will live to reach maturity.

Other animals, like reptiles, amphibians, and fish have very little chance of surviving to reaching maturity.

A crocodile lays up to about 90 eggs. Most of them never hatch because predators eat them. Fewer than 1 out of 10 of the baby crocodiles will live to become adult crocodiles. ➤

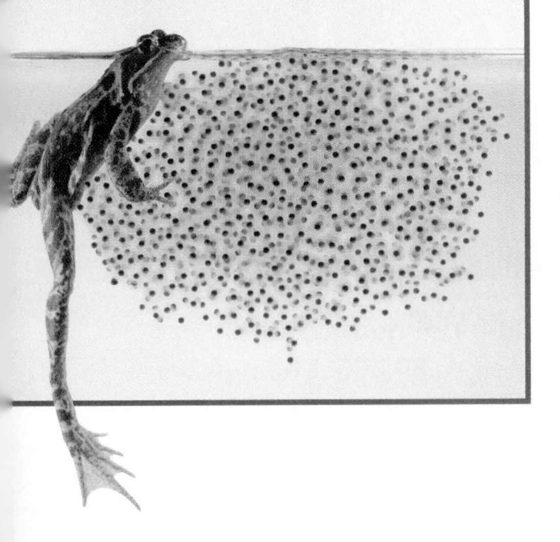

◄ Some frogs can lay 1,000 or more eggs. Fewer than 1 out of 100 of the tadpoles that hatch from the eggs will live to become adult frogs.

Many fish have even less chance of survival. The walleye lays more than 200,000 eggs. Usually, fewer than 1 out of 1,000 of the eggs will become adult walleye. ➤

Patterns of Predators and Prey

In 1995, twenty-five gray wolves were brought to Yellowstone National Park, where no wolves had lived for over 70 years. Since wolves eat elk, scientists wondered what would happen to the elk population.

When the wolves were introduced, there were 170,000 elk living at the park. ▼

▲ By 2004, the number of elk had dropped to 80,000. The number of wolves had increased 10 times to 250.

Scientists want the wolf population to grow, but they don't want the elk population to shrink too much. They work to find and keep a healthy balance of wolves and elk in the park.

What other patterns in the wild do you think scientists might study?

Data Bank

Drinks Vending Machine Poster

Example The "Exact Change" light is on and you want to buy apple juice.

- Use any combination of coins that equals 55 cents.

The "Exact Change" light is not on and you want to buy apple juice.

- If you use a dollar bill, the machine will give you the apple juice and 45 cents in change.
- If you use 3 quarters, the machine will give you the apple juice and 20 cents in change.
- If you use 2 quarters and 1 dime, the machine will give you the apple juice and 5 cents in change.

Snacks Vending Machine Poster

Stationery Store Poster

$1.79

SALE! Reg. $1.89

Correction Fluid

$2.99

SALE! Reg. $4.49

Pens
Box of 24

Photo Album

$3 off

99¢

SALE! Reg. $1.39

Paper Clips
Box of 100

$1.49

SALE!
Reg. $1.89

Pencils
Box of 24

$2.99

SALE! Reg. $5.19

Batteries
8-Pack

99¢

SALE! Reg. $1.89

Crayons
Box of 16

BUY-RITE COUPON

49¢

Without coupon 99¢

Notebook

Variety Store Poster

Toys

- **Mini stock cars**
  10 per box **$2.99** per box
- **Marbles**
  45 per bag **$1.45** per bag
- **Interlocking building blocks**
  395 pieces **$19.99** per set

Fashion

- **Bright shoelaces**
  5 pairs per package **$2.99** per pkg.
- **Ponytail rings**
  12 per package **$1.77** per pkg.
- **"Hair Things"**
  6 per bag **$1.00** per bag

School Supplies

- **Notebook paper**
  200 sheets per package **$0.98** per pkg.
- **Value Pack pens**
  10 in a package **$1.27** per pkg.
- **Chocolate-scented pens!**
  6 in a pack **$1.29** per pack
- **"Fashion" pens**
  4 in a pack **$1.29** per pack
- **File Cards**

  package of 100 for **$1.69**
- **Brilliant color markers**
  5-pack **$1.99**
- **Scented markers**
  package of 8 for **$2.69**
- **Pencils**
  8-pack **$1.00** *6-pack* **$0.69**

Party Supplies

- **Glitter Stickers**
  7 per pack **$1.00** per pack
- **9-inch balloons**
  25 per bag **$1.99** per bag
- **Party hats** *6 for* **$1.49**
  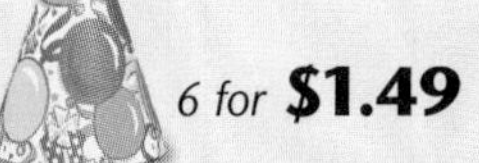
- **Party horns** *8 for* **$2.99**
- **Giant 14-inch balloons**
  package of 5 for **$1.79**

Stock-Up Sale Poster #1

Light Bulbs
4-Pack **$1.09**

5 or More Sale — *You pay $0.88 per pack*

Extension Cord
$3.25

5 or More Sale — *You pay $2.79 per cord*

Tissues
$0.73

5 or More Sale — *You pay $0.57 per box*

Transparent Tape
$0.84

5 or More Sale — *You pay $0.65 per roll*

Batteries
4-Pack **$3.59**

5 or More Sale — *You pay $2.90 per pack*

Toothpaste
$1.39

5 or More Sale — *You pay $1.14 per tube*

Ballpoint Pen
$0.39

5 or More Sale — *You pay $0.27 per pen*

Tennis Balls
Can of 3 **$2.59**

5 or More Sale — *You pay $1.86 per can*

Paperback Book
$2.99

5 or More Sale — *You pay $2.25 per book*

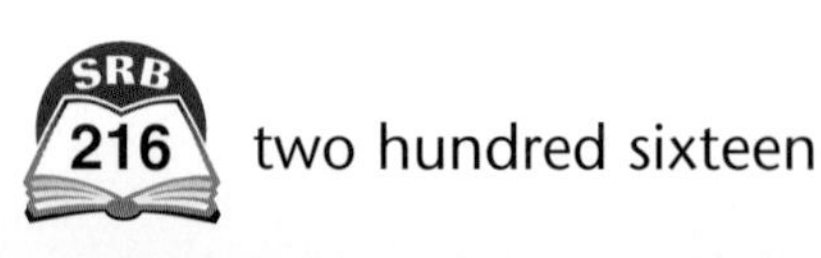

Stock-Up Sale Poster #2

Greeting Cards
Box of 12 **$3.29**

5 or More Sale *You pay $2.63 per box*

Bath Soap
$0.88

5 or More Sale *You pay $0.65 per bar*

Gift Wrapping Paper
$2.35 per roll

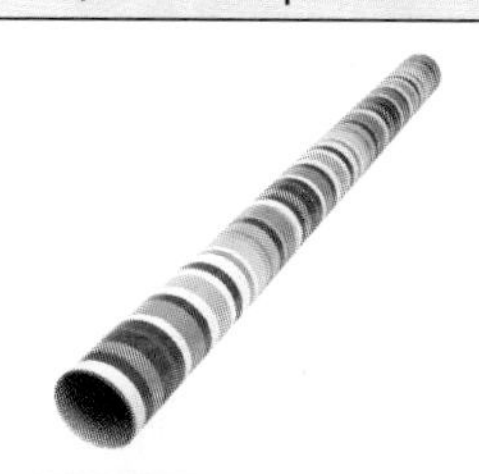

5 or More Sale *You pay $1.86 per roll*

Toothbrush
$1.38

5 or More Sale *You pay $1.13 per brush*

Garbage Bags
$3.75

5 or More Sale *You pay $3.18 per box*

Night Light Bulbs
2-Pack **$0.96**

5 or More Sale *You pay $0.76 per pack*

Glue
$1.15

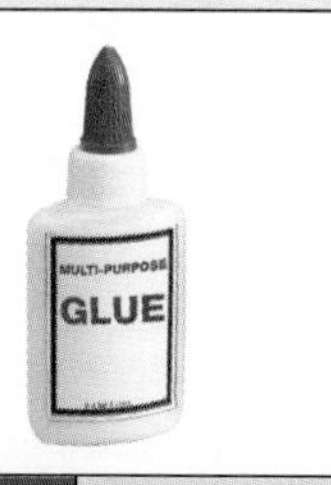

5 or More Sale *You pay $0.94 per bottle*

Construction Paper
$0.67 per pad

5 or More Sale *You pay $0.54 per pad*

Shoelaces
$1.27 per pair

5 or More Sale *You pay $1.08 per pair*

Animal Clutches

All of the animals shown lay eggs. A nest of eggs is called a *clutch.*

Most birds, reptiles, and amphibians lay eggs once or twice a year. Insects may lay eggs daily during a certain season of the year.

Green Turtle

up to 1.5 meters long

median of 104 eggs,
as many as 184 eggs

Giant Toad

up to 30 cm long

maximum of 35,000 eggs

Ostrich

more than 2 meters tall

up to 15 eggs

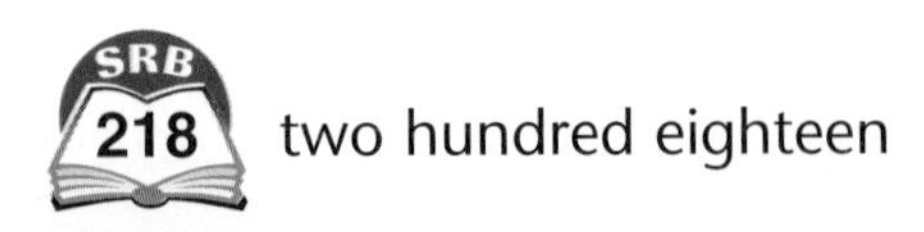

Python

up to 9 meters long

median of 29 eggs, as many as 100

Agama lizard

up to 25 cm long

up to 23 eggs

Queen termite

less than 1 cm long

as many as 8,000 eggs per day for years

Mississippi alligator

up to 4.5 meters long

as many as 88 eggs recorded

Normal Spring High and Low Temperatures (in °F)

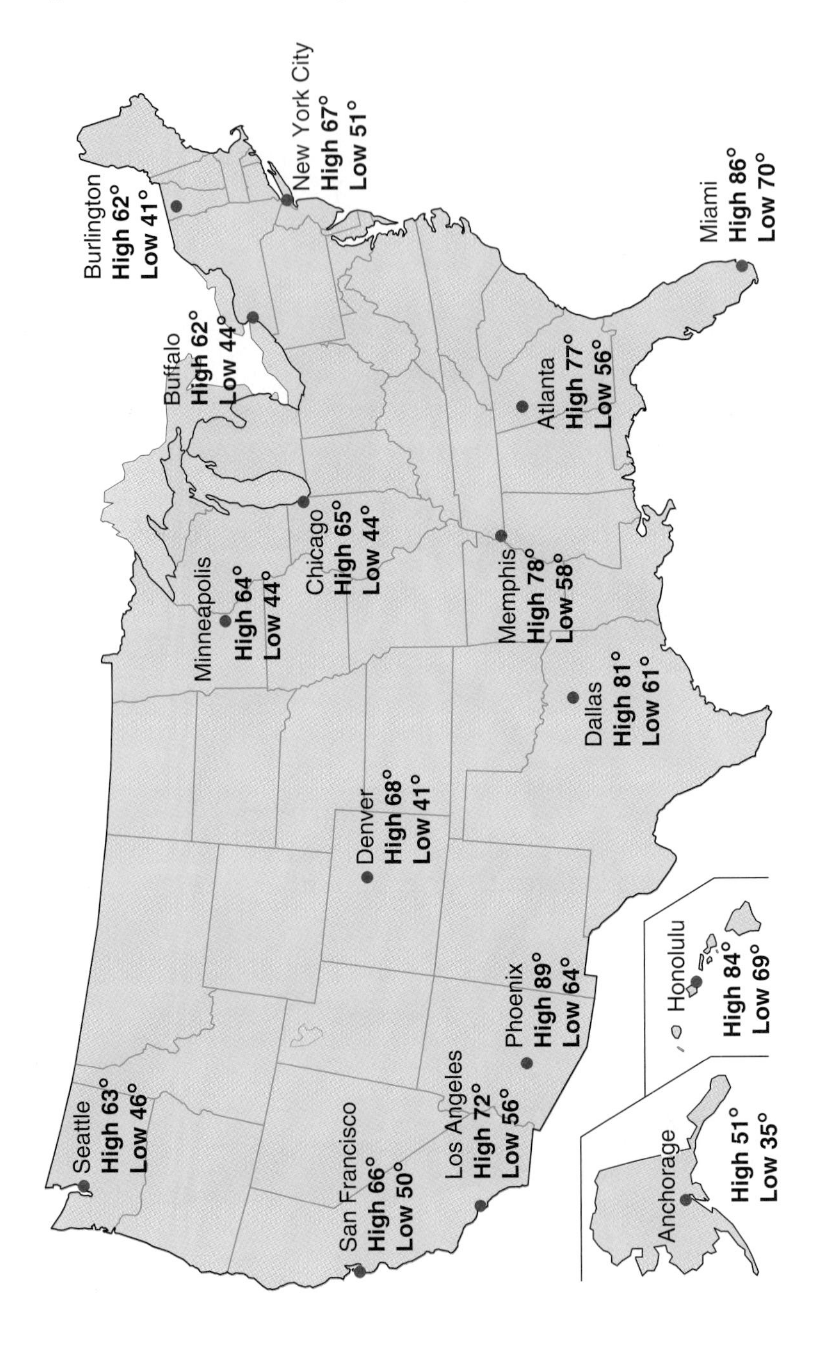

Temperatures given are averages for 30 years of data.
70°F is normal room temperature.

Normal September Rainfall (in centimeters)

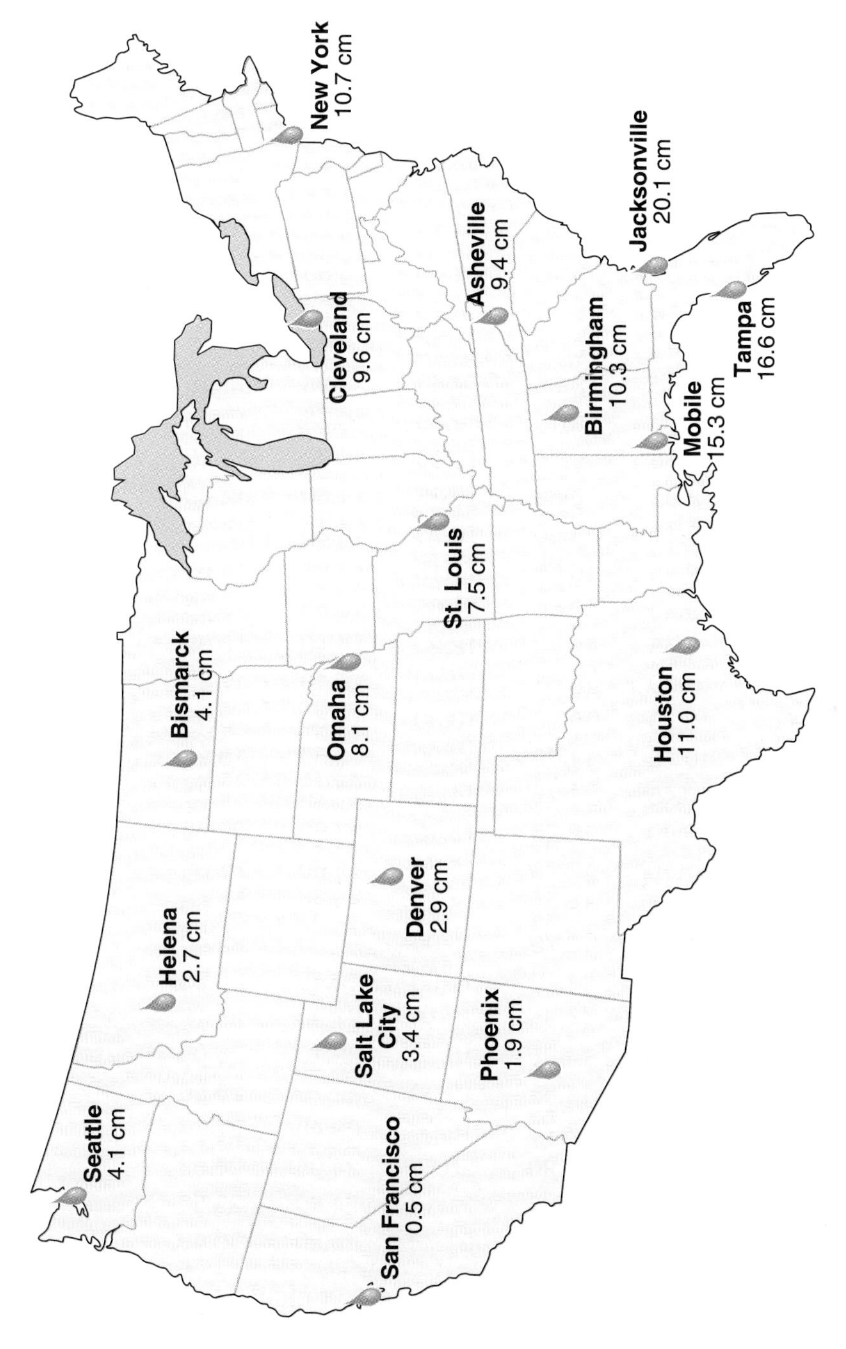

Amounts given are averages for 30 years of data.

Shipping Packages: Rate Table

Ground Service							
Residential Deliveries (Delivery to a home)							
WEIGHT NOT TO EXCEED	ZONES						
	2	3	4	5	6	7	8
1 lb	$6.25	$6.40	$6.75	$6.85	$7.15	$7.20	$7.35
2	6.35	6.65	7.20	7.35	7.80	7.95	8.35
3	6.50	6.85	7.50	7.75	8.20	8.45	9.10
4	6.65	7.15	7.85	8.15	8.60	8.85	9.60
5	6.90	7.35	8.10	8.45	8.95	9.25	10.10
6	7.05	7.50	8.25	8.70	9.20	9.60	10.40
7	7.30	7.65	8.45	8.95	9.50	9.95	10.80
8	7.50	7.85	8.60	9.15	9.80	10.35	11.45
9	7.70	8.05	8.80	9.35	10.00	10.80	12.10
10	7.90	8.25	8.95	9.60	10.35	11.50	12.85
11	8.10	8.40	9.10	9.80	10.75	12.20	13.70
12	8.35	8.65	9.25	10.00	11.15	12.95	14.60
13	8.55	8.90	9.40	10.20	11.65	13.65	15.45
14	8.70	9.10	9.55	10.40	12.20	14.40	16.35
15	8.90	9.35	9.70	10.65	12.80	15.10	17.25
16	9.00	9.60	9.90	10.90	13.30	15.80	18.05
17	9.10	9.85	10.15	11.30	13.90	16.50	18.95
18	9.25	10.10	10.45	11.75	14.50	17.20	19.85
19	9.40	10.40	10.80	12.25	15.10	17.90	20.75
20	9.60	10.70	11.10	12.75	15.70	18.60	21.65
21	9.80	11.00	11.45	13.25	16.30	19.30	22.50
22	10.00	11.30	11.80	13.70	16.90	20.00	23.40
23	10.20	11.60	12.10	14.15	17.50	20.65	24.30
24	10.40	11.85	12.45	14.55	18.10	21.40	25.15
25	10.60	12.10	12.80	14.95	18.70	22.10	26.05
26	10.85	12.40	13.15	15.35	19.30	22.75	26.90
27	11.05	12.60	13.50	15.75	19.90	23.40	27.75
28	11.25	12.85	13.85	16.20	20.50	24.05	28.65
29	11.45	13.15	14.20	16.65	21.10	24.80	29.50
30	11.70	13.40	14.55	17.10	21.70	25.50	30.40

FOR ANY FRACTION OF A POUND OVER THE WEIGHT SHOWN, USE THE NEXT HIGHER RATE.

Shipping Packages: Zone Map

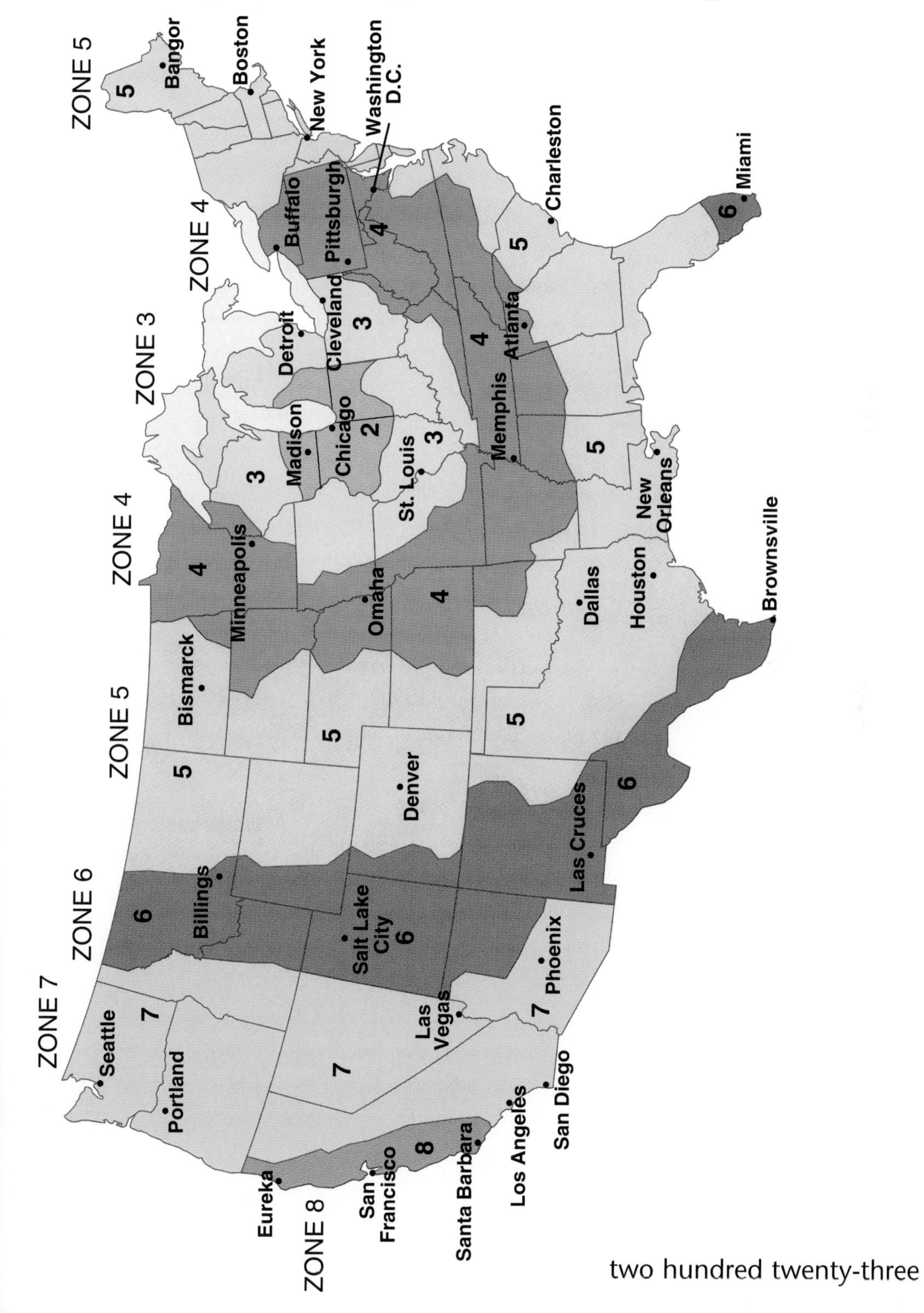

U.S. Road Mileage Map

Except for Hawaii and Alaska, all numbers are highway distances in miles.

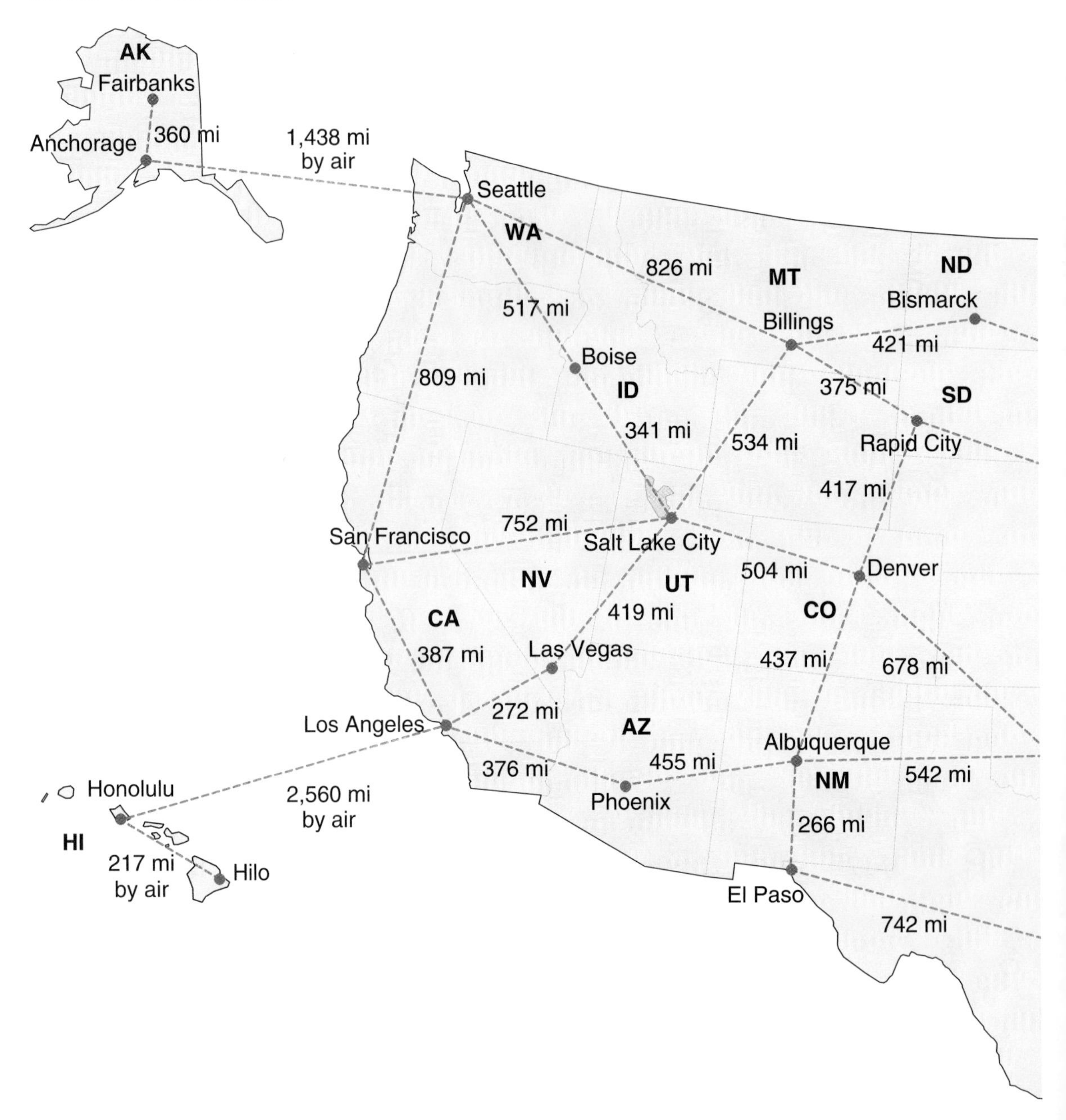

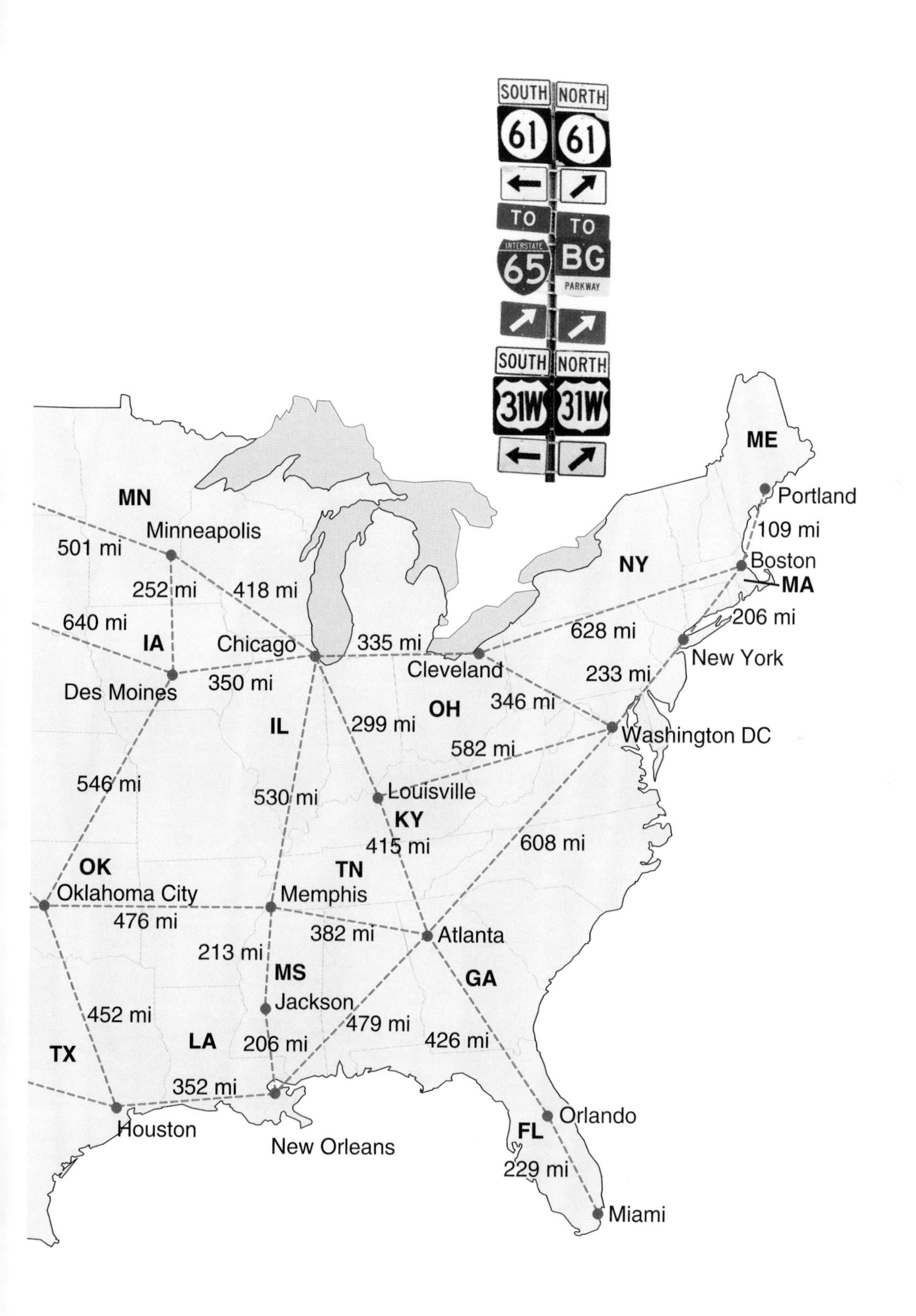
SOUTH
NORTH
61
61
TO
TO
INTERSTATE
65
BG
PARKWAY
SOUTH
NORTH
31W
31W
MN
Minneapolis
501 mi
252 mi
418 mi
640 mi
IA
Chicago
335 mi
Des Moines
350 mi
Cleveland
IL
299 mi
OH
346 mi
582 mi
546 mi
530 mi
Louisville
KY
415 mi
608 mi
OK
Oklahoma City
476 mi
TN
Memphis
382 mi
Atlanta
213 mi
MS
GA
Jackson
452 mi
479 mi
426 mi
TX
LA
206 mi
352 mi
Houston
New Orleans
Orlando
FL
229 mi
Miami
ME
Portland
109 mi
Boston
MA
NY
206 mi
628 mi
New York
233 mi
Washington DC

Major U.S. City Populations

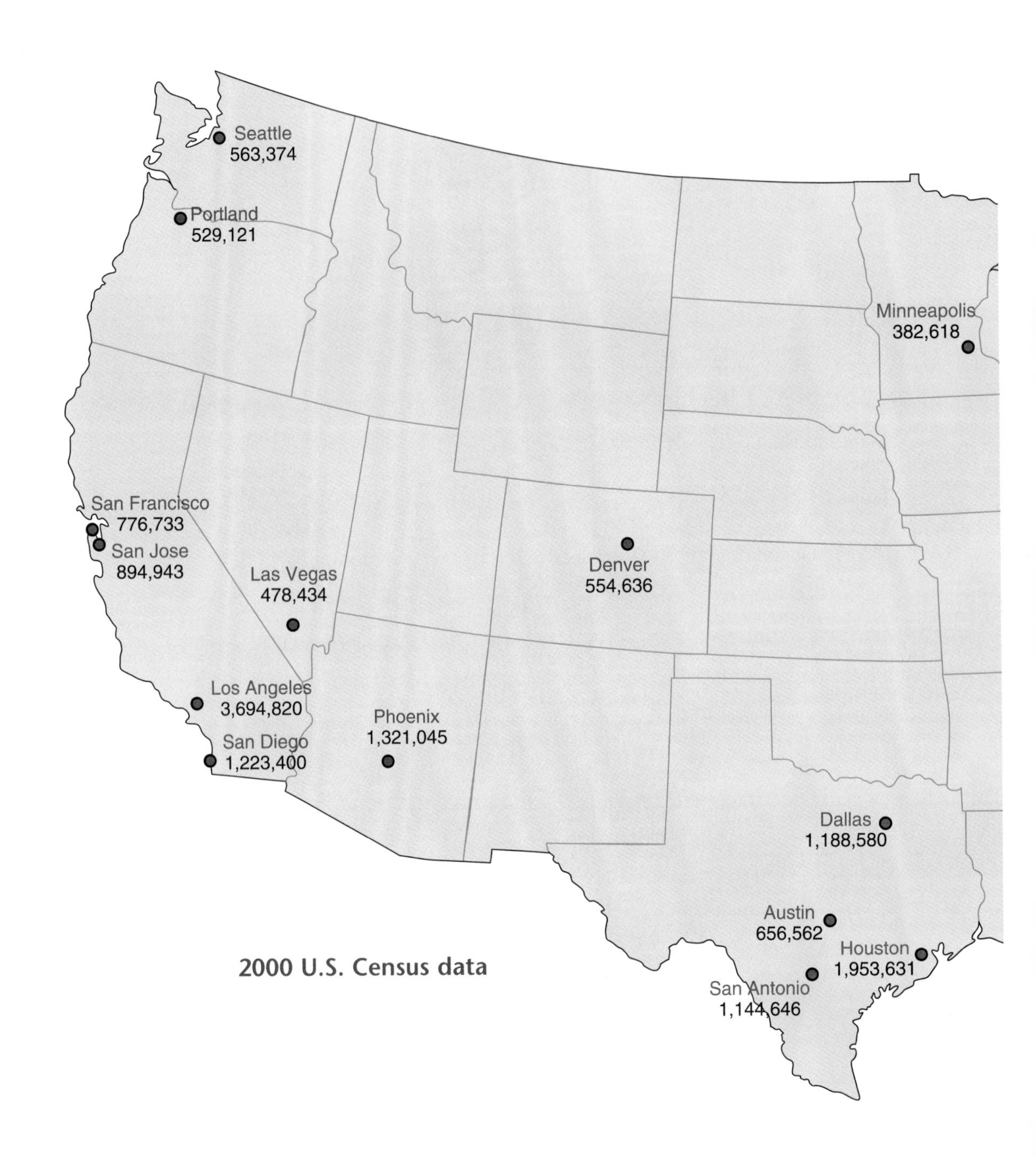

2000 U.S. Census data

Boston
589,141
Milwaukee
596,974
Detroit
951,270
New York
8,008,278
Philadelphia
1,517,550
Chicago
2,896,016
Columbus
711,470
Baltimore
651,154
Indianapolis
791,926
Washington, D.C.
572,059
St. Louis
348,189
Charlotte
540,828
Memphis
650,100
Atlanta
416,474
Jacksonville
735,617
New Orleans
484,674
Miami
362,470

Sizes of Sport Balls

Most sport balls are spheres. The size of a sphere is the distance across the center of the sphere. This distance is called the **diameter of the sphere.**

The table lists diameters for sport balls that are spheres. Each diameter is given in inches and in centimeters.

The segment *RS* passes through the center of the sphere. The length of this segment is the diameter of the sphere.

Diameters of Sport Balls

Ball	Diameter in Inches	Diameter in Centimeters
Table tennis	$1\frac{1}{2}$ in.	3.8 cm
Squash	$1\frac{5}{8}$ in.	4.1 cm
Golf	$1\frac{11}{16}$ in.	4.3 cm
Tennis	$2\frac{5}{8}$ in.	6.7 cm
Baseball	$2\frac{3}{4}$ in.	7.0 cm
Cricket	$2\frac{13}{16}$ in.	7.1 cm
Croquet	$3\frac{5}{8}$ in.	9.2 cm
Softball	$3\frac{13}{16}$ in.	9.7 cm
Volleyball	$8\frac{1}{4}$ in.	21.0 cm
Bowling	$8\frac{1}{2}$ in.	21.6 cm
Soccer	$8\frac{3}{4}$ in.	22.2 cm
Water polo	$8\frac{7}{8}$ in.	22.5 cm
Basketball	$9\frac{1}{2}$ in.	24.1 cm

Reminder:

1 inch is about 2.5 centimeters

1 centimeter is about 0.4 inch

Weights of Sport Balls

The table below lists weights for sport balls. Each weight is given in ounces and in grams.

Weights of Sport Balls

Ball	Weight in Ounces	Weight in Grams
Table tennis	$\frac{1}{10}$ oz	2.5 g
Squash	$\frac{9}{10}$ oz	25 g
Golf	$1\frac{1}{2}$ oz	43 g
Tennis	2 oz	57 g
Baseball	5 oz	142 g
Cricket	$5\frac{1}{2}$ oz	156 g
Softball	$6\frac{1}{2}$ oz	184 g
Volleyball	$9\frac{1}{2}$ oz	270 g
Soccer	15 oz	425 g
Water polo	15 oz	425 g
Croquet	16 oz	454 g
Basketball	22 oz	625 g
Bowling	256 oz	7,260 g

Reminder:

16 ounces equals 1 pound

1,000 grams equals 1 kilogram

1 ounce is about 28 grams

1 kilogram is about 2.2 pounds

Check Your Understanding

1. Which sport balls weigh 1 pound or more?
2. Which sport balls weigh 1 kilogram or more?
3. Do the heaviest sport balls also have the largest diameters?

Check your answers on page 343.

Physical Fitness Standards

The table on the next page shows data for three fitness tests.

Curl-Ups

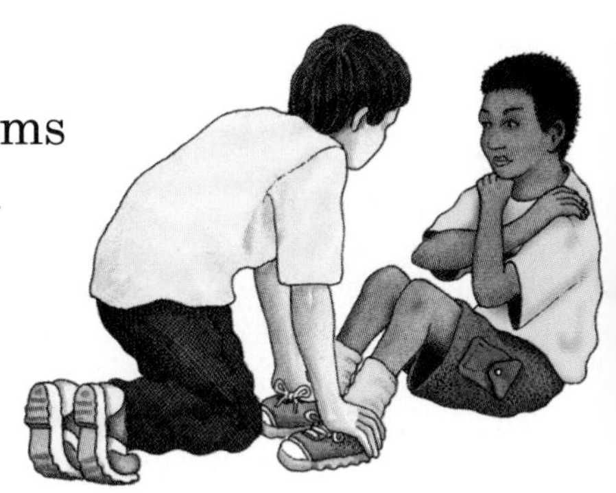

A partner holds your feet. You cross your arms and place your hands on opposite shoulders. You raise your body and curl up to touch your elbows to your thighs. Then you lower your back to the floor. This counts as one curl-up. Do as many curl-ups as you can in one minute.

One Mile Run/Walk

You cover a 1-mile distance in as short a time as you can. You may not be able to run the entire distance. Walk when you are not able to run.

Arm Hang

Hold the bar with your palms facing away from your body. Your chin should clear the bar. (See the picture.) Hold this position as long as you can.

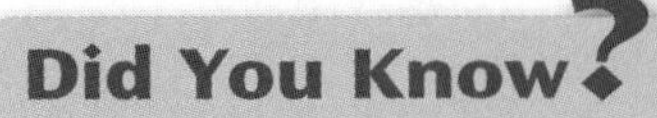

In order to be physically fit, children between the ages of 7 and 11 need at least 9 hours of sleep each night.

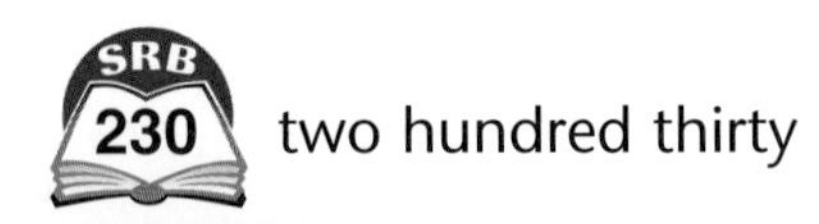

Physical Fitness Test Scores (median scores for each age)				
	Age	Curl-Ups (in 1 minute)	1-Mile Run (minutes:seconds)	Arm Hang (seconds)
	6	22	12:36	6
B	7	28	11:40	8
O	8	31	11:05	10
Y	9	32	10:30	10
S	10	35	9:48	12
	11	37	9:20	11
	12	40	8:40	12
	6	23	13:12	5
G	7	25	12:56	6
I	8	29	12:30	8
R	9	30	11:52	8
L	10	30	11:22	8
S	11	32	11:17	7
	12	35	11:05	7

Source: The National Physical Fitness Award: Qualifying Standards

Example The table shows 31 curl-ups for 8-year-old boys. 31 curl-ups is the *median* score for 8-year-old boys. About half of all 8-year-old boys will do *31 or more* curl-ups, and about half of all 8-year-old boys will do *31 or fewer* curl-ups.

Example The table shows a *median* time of 12 minutes and 30 seconds for 8-year-old girls in the mile run. About half of all 8-year-old girls will take *12:30 or longer* to run a mile, and about half of all 8-year-old girls will take *12:30 or less.*

Record High and Low Temperatures

The table shows the highest and lowest temperatures ever recorded in each state.

State Record Temperatures (in degrees Fahrenheit)

State	Lowest °F	Highest °F	State	Lowest °F	Highest °F
Alabama	−27	112	Missouri	−40	118
Alaska	−80	100	Montana	−70	117
Arizona	−40	128	Nebraska	−47	118
Arkansas	−29	120	Nevada	−50	125
California	−45	134	New Hampshire	−47	106
Colorado	−61	118	New Jersey	−34	110
Connecticut	−32	106	New Mexico	−50	122
Delaware	−17	110	New York	−52	108
			N. Carolina	−34	110
District of Columbia	−15	106	N. Dakota	−60	121
Florida	−2	109	Ohio	−39	113
Georgia	−17	112	Oklahoma	−27	120
Hawaii	12	100	Oregon	−54	119
Idaho	−60	118	Pennsylvania	−42	111
Illinois	−36	117	Rhode Island	−25	104
Indiana	−36	116	S. Carolina	−19	111
Iowa	−47	118	S. Dakota	−58	120
Kansas	−40	121	Tennessee	−32	113
Kentucky	−37	114	Texas	−23	120
Louisiana	−16	114	Utah	−50	117
Maine	−48	105	Vermont	−69	105
Maryland	−40	109	Virginia	−30	110
Massachusetts	−35	107	Washington	−48	118
Michigan	−51	112	W. Virginia	−37	112
Minnesota	−60	114	Wisconsin	−55	114
Mississippi	−19	115	Wyoming	−66	115

Tornado Data

Tornadoes are violent storms. A tornado looks like a narrow black cloud that is shaped like a funnel. The funnel reaches down toward the ground. The tip of the funnel touches the ground as it moves along.

The winds of a tornado can reach speeds of 500 miles per hour.

Nearly all tornadoes occur in the United States. Tornadoes are very common in some states. But other states hardly ever have tornadoes.

Tornado Map for 1995–2000

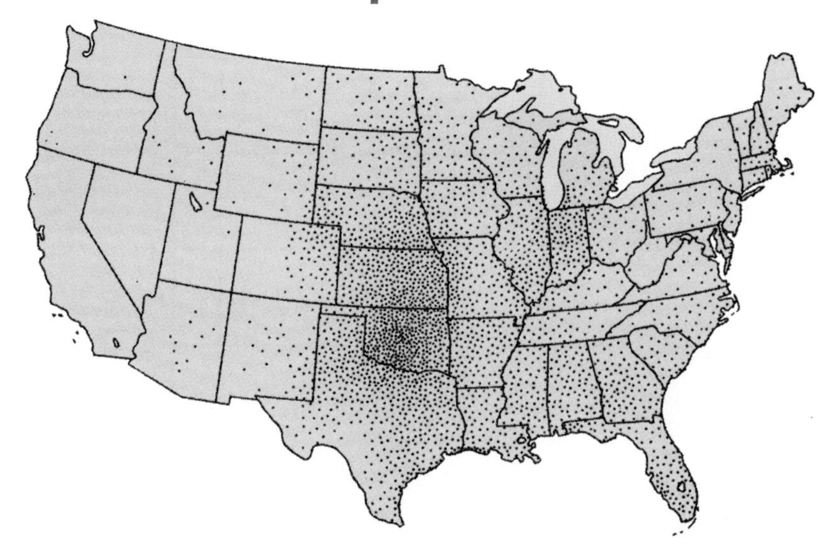

Each tornado that occurred between 1995 and 2000 is shown by a dot. The dot marks the spot where the tornado first touched the ground.

World Population Growth

There are more than 6 billion people in the world today. The table and graph below show how the world's population has grown.

By the year 2050, the world will have about 10 billion people.

World Population Table

Date	Population	Date	Population
1000 B.C.	50,000,000	1927	2,000,000,000
A.D. 1	300,000,000	1960	3,000,000,000
1250	400,000,000	1974	4,000,000,000
1500	500,000,000	1987	5,000,000,000
1804	1,000,000,000	1999	6,000,000,000

Source: United Nations Population Division

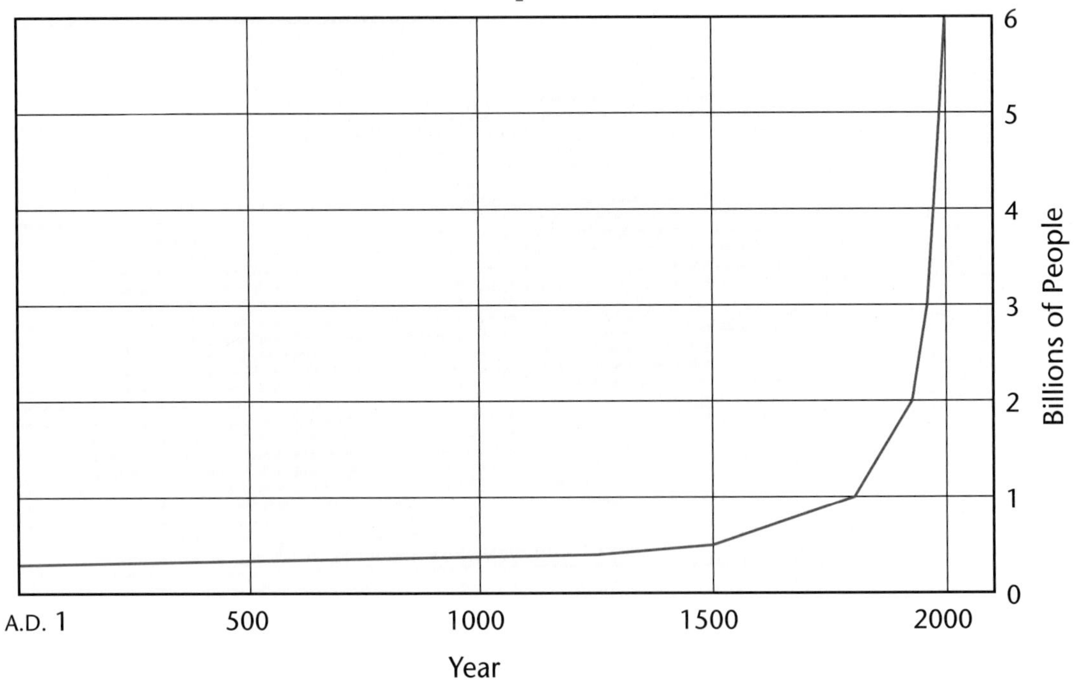

Heights of 8-Year-Old Children

The table below shows heights that were measured to the nearest centimeter. All of the boys and girls were 8 years old.

Heights of Third Graders			
Boys		Girls	
Boy	Height	Girl	Height
#1	136 cm	#1	123 cm
#2	129 cm	#2	141 cm
#3	110 cm	#3	115 cm
#4	122 cm	#4	126 cm
#5	126 cm	#5	122 cm
#6	148 cm	#6	144 cm
#7	127 cm	#7	127 cm
#8	126 cm	#8	133 cm
#9	124 cm	#9	120 cm
#10	142 cm	#10	125 cm
#11	118 cm	#11	126 cm
#12	130 cm	#12	107 cm

The average 8-year-old boy is slightly taller than the average 8-year-old girl. How can you use the data in the table to show this?

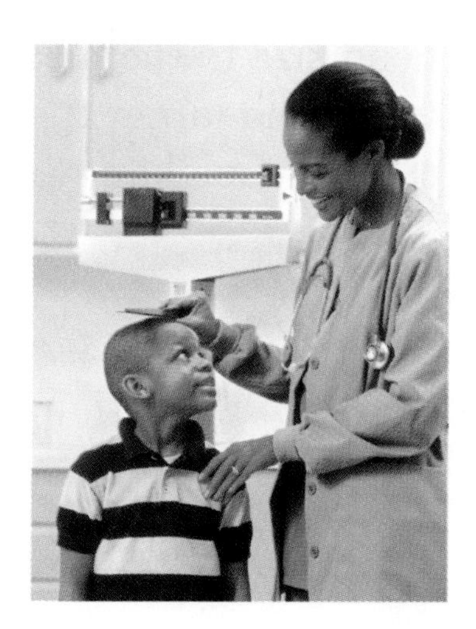

Head Size

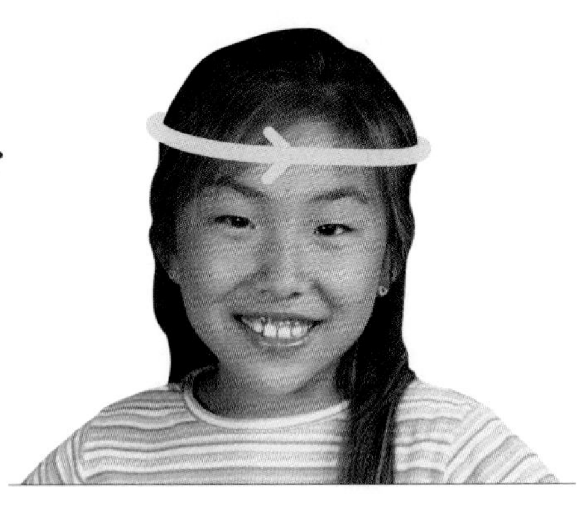

Your head size is the distance around your head. You can use a tape measure to measure the distance around. The line graph on the next page shows how head size increases as you get older. The graph shows the *median* head size for each age.

Example The graph shows a *median* head size of 52 centimeters for an 8-year-old. About half of all 8-year-olds have a head size *larger* than 52 cm. And about half have a head size *smaller* than 52 cm.

Describe what happens to a baby's head size during the first year of its life. Is your head growing as fast as a baby's head?

Ask someone to measure your head size. Use the centimeter scale of the tape measure. Compare your head size to the median head size for your age. Is your head larger than the median head size or smaller than the median head size?

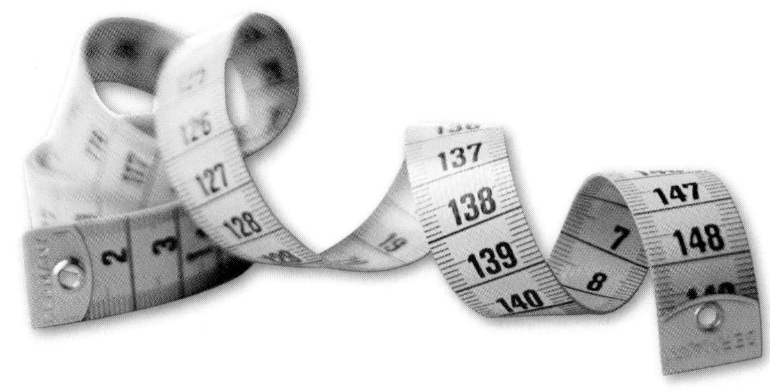

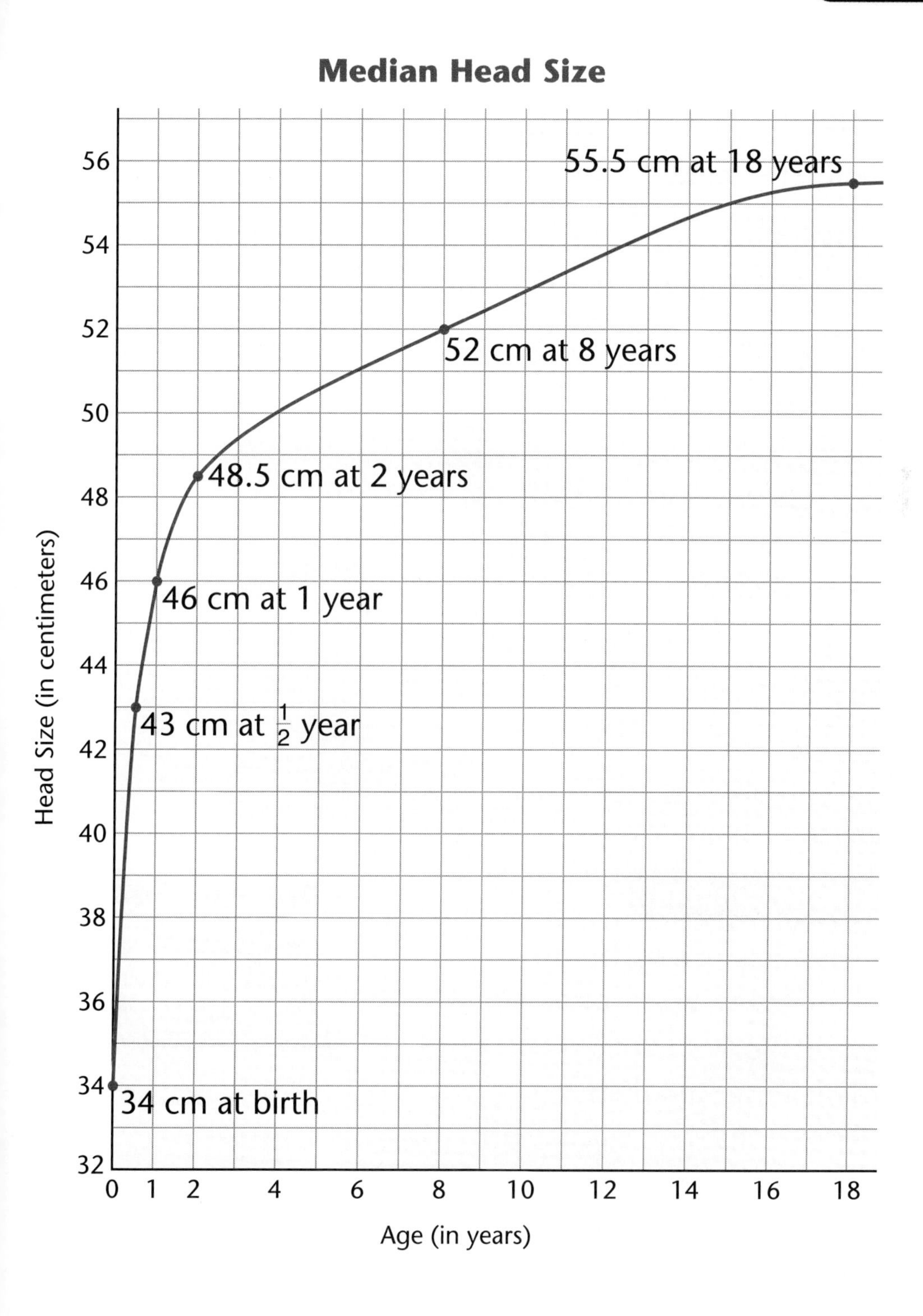
Median Head Size
55.5 cm at 18 years
52 cm at 8 years
48.5 cm at 2 years
46 cm at 1 year
43 cm at $\frac{1}{2}$ year
34 cm at birth
Head Size (in centimeters)
56
54
52
50
48
46
44
42
40
38
36
34
32
0 1 2 4 6 8 10 12 14 16 18
Age (in years)

Number of Words Children Know

When babies are about one year old, they begin to imitate sounds. They understand some words.

By the time a child is 6 years old, he or she understands and uses between 2,000 and 3,000 words.

The table below gives the average number of words that children understand and use.

Number of Words Children Understand and Use

Age in Years	Number of Words	Age in Years	Number of Words
1	3	$3\frac{1}{2}$	1,222
$1\frac{1}{2}$	22	4	1,540
2	272	$4\frac{1}{2}$	1,870
$2\frac{1}{2}$	446	5	2,072
3	896	6	2,562

Check Your Understanding

1. Today is Judy's 5th birthday. About how many *new* words will she use during the next year?
2. Today is Isaac's 2nd birthday. About how many *new* words will he use during the next year?

Check your answers on page 343.

Did You Know?

The largest English-language dictionary lists about 300,000 different words.

Letter Frequencies

There are 26 letters in the English alphabet. Some letters (like E and T) are used quite often. Other letters (like Q and Z) are not used very much.

The table below shows how often each letter is used in writing English words. If you looked at 1,000 letters, you could expect to see about this number of each letter.

Letter Frequencies	
82 **A**s	70 **N**s
14 **B**s	80 **O**s
28 **C**s	20 **P**s
38 **D**s	1 **Q**
130 **E**s	68 **R**s
30 **F**s	60 **S**s
20 **G**s	105 **T**s
53 **H**s	25 **U**s
65 **I**s	9 **V**s
1 **J**	15 **W**s
4 **K**s	2 **X**s
34 **L**s	20 **Y**s
25 **M**s	1 **Z**

Did You Know?

Scrabble is a word game that uses 100 tiles. A letter is printed on each tile, and the tiles are used to form words. Vowels (the letters A, E, I, O, and U) are printed on 42 of these tiles.

Check Your Understanding

1. Which 5 letters are used the most?
2. Which 5 letters are used the least?
3. The letters A, E, I, O, and U are called *vowels*. How many vowels would you expect to see if you looked at 1,000 letters?

Check your answers on page 343.

Heights and Depths

The table below shows the highest point on each continent. Each height is a distance (in feet) above sea level.

Continent	Highest Point	Height (feet)
Asia	Mount Everest, Nepal/Tibet	29,028
South America	Mount Aconcagua, Argentina	22,834
North America	Mount McKinley, U.S.A.	20,320
Africa	Mount Kilimanjaro, Tanzania	19,340
Antarctica	Vinson Massif, Sentinal Rouge	16,864
Europe	Mont Blanc, France/Italy	15,771
Australia	Mount Kosciusko, New South Wales	7,310

The table to the right shows the greatest depth for each ocean. A negative sign (−) is used to show that each depth is a distance below sea level.

Ocean	Greatest Depth (feet)
Pacific	−36,200
Atlantic	−30,246
Indian	−24,442
Arctic	−17,881

The table below shows distance above sea level for several high and low places in the United States. Negative heights (−) are distances below sea level.

Location	Height (feet)
Mt. Whitney, California	14,494
Death Valley, California	−282
Mauna Kea, Hawaii	13,796
Highest point in Louisiana	535
New Orleans, Louisiana	−8
Mount Katahdin, Maine	5,267
Mount Hood, Oregon	11,239
Highest point in S. Dakota	7,242

Mount Hood

How Much Would You Weigh on the Moon?

All objects are attracted to each other by a force known as "gravity." Gravity pulls your body toward the center of the Earth.

When you weigh yourself, you step on a scale. The reading on the scale measures the pull of Earth's gravity on your body.

The pull of the moon's gravity is much less than the pull of Earth's gravity. So you would weigh much less on the moon. The table below shows how a person's weight would change if they could travel to the planets, sun, and moon. The weights are for a space traveler who weighs 100 pounds on Earth.

Weight Changes for a Space Traveler

Location	Person's Weight
Earth	100 pounds
Moon	17 pounds
Sun	2,800 pounds
Mercury	38 pounds
Venus	90 pounds
Mars	38 pounds
Jupiter	236 pounds
Saturn	92 pounds
Uranus	89 pounds
Neptune	112 pounds

Ages of U.S. Presidents

The table below gives the age of each man when he became president.

President	Age	President	Age
1. George Washington	57	23. Benjamin Harrison	55
2. John Adams	61	24. Grover Cleveland	55
3. Thomas Jefferson	57	25. William McKinley	54
4. James Madison	57	26. Theodore Roosevelt	42
5. James Monroe	58	27. William Taft	51
6. John Quincy Adams	57	28. Woodrow Wilson	56
7. Andrew Jackson	61	29. Warren G. Harding	55
8. Martin Van Buren	54	30. Calvin Coolidge	51
9. William Harrison	68	31. Herbert Hoover	54
10. John Tyler	51	32. Franklin D. Roosevelt	51
11. James Polk	49	33. Harry S. Truman	60
12. Zachary Taylor	64	34. Dwight D. Eisenhower	62
13. Millard Fillmore	50	35. John F. Kennedy	43
14. Franklin Pierce	48	36. Lyndon B. Johnson	55
15. James Buchanan	65	37. Richard Nixon	56
16. Abraham Lincoln	52	38. Gerald Ford	61
17. Andrew Johnson	56	39. James Carter	52
18. Ulysses S. Grant	46	40. Ronald Reagan	69
19. Rutherford B. Hayes	54	41. George H. W. Bush	64
20. James Garfield	49	42. William Clinton	46
21. Chester Arthur	51*	43. George W. Bush	54
22. Grover Cleveland	47		

*Some resources list Chester Arthur as having been born October 5, 1829; others say 1830. He became president (after the death of Garfield) on September 20, 1881. At that time he was either 50 or 51 years old.

Did You Know?

Unless the previous president dies or resigns, a president is sworn in and takes office on January 20, several months after the presidential election. Before 1937, the president took office on March 4.

Theodore Roosevelt in carriage on way to being sworn in March 4, 1905.

Check Your Understanding

1. Who was the youngest person to become president?

2. Who was the oldest person?

Check your answers on page 343.

Train Schedule and Airline Schedule

Train Schedule for a Chicago Rail Line

Station	Time
South Chicago	11:46 A.M.
83rd Street	11:49
Cheltenham	11:51
South Shore	11:55
Bryn Mawr	11:57
59th Street	**12:04 P.M.**
Hyde Park	**12:08**
Kenwood	**12:09**
McCormick Place	**12:14**
18th Street	**12:15**
Van Buren Street	**12:19**
Randolph Street	**12:22**

Airline Schedule from Chicago to New York

Departure	Arrival
6:00 A.M.	8:59 A.M.
6:20 A.M	1:12 P.M.*
7:00 A.M	9:56 A.M.
7:00 A.M	10:04 A.M.
8:00 A.M	11:00 A.M.
8:45 A.M	2:00 P.M.*
9:00 A.M	12:00 P.M.
10:00 A.M	12:58 P.M.
10:20 A.M.	3:19 P.M.*
11:00 A.M	1:55 P.M.
12:00 P.M.	3:00 P.M.
1:00 P.M.	3:55 P.M.
1:20 P.M.	4:21 P.M.
1:20 P.M.	6:45 P.M.*
1:30 P.M.	4:39 P.M.
2:00 P.M.	5:09 P.M.
3:00 P.M.	6:04 P.M.
4:00 P.M.	7:00 P.M.
4:14 P.M.	9:19 P.M.*
4:40 P.M.	7:30 P.M.
5:00 P.M.	8:01 P.M.
6:00 P.M.	9:02 P.M.
7:00 P.M.	10:00 P.M.

*Flight makes other stops.

Times shown are local times.

New York time is 1 hour ahead of Chicago time.

More Information About North American Animals

North America is the 3rd largest continent.

- Asia and Africa are larger.
- South America, Antarctica, Europe, and Australia are smaller.

The area of North America is about 9,400,000 square miles.

- It contains $\frac{1}{6}$ of the world's *land* area.
- It contains $\frac{1}{20}$ of the world's *total* area. Total area equals land area plus water area.

North America is $2\frac{1}{2}$ times larger than the United States.

Although the animals on journal pages 204 and 205 are positioned on or near areas they inhabit, their natural habitats may greatly exceed the locations in which they are shown. Here is more information about the ranges of these animals.

American alligator
Coastal areas from North Carolina to Florida, and along the Gulf states from Florida to Texas

American porcupine
Canada and Alaska, except arctic regions; U.S., except the southeastern and Gulf states

Arctic fox
Arctic regions and northern Canada

Atlantic green turtle (endangered species)
Warm tropical waters of the Atlantic Ocean, Gulf of Mexico, and Caribbean Sea; from Texas to Massachusetts along the U.S. coast.

Beaver
Canada and Alaska, except arctic regions; U.S., except southwestern states; northeastern Mexico

Beluga whale
Arctic and sub-arctic coasts of North America, Asia, and Europe

Black bear
Canada and Alaska, except arctic regions; Northern U.S., and Florida to Louisiana; Appalachian and Rocky Mountains, extending into northern Mexico

Bottle-nose dolphin
Temperate and tropical waters throughout the world's oceans

Common dolphin
Temperate and tropical coastlines throughout the world; Atlantic coast from Newfoundland to Florida; Pacific coast and open water from Washington State to Chile

Gila monster
Arizona and small areas of bordering states; Northern Mexico

Gray fox
U.S., except northwestern and north-central states; Mexico to northern South America

Gray whale
Pacific coast, from Alaska to Mexico

Harp seal
Arctic waters of the Atlantic Ocean, as far south as Maine

Mountain goat
Mountainous regions in Canada, southern Alaska, and northwestern U.S.

Northern fur seal
Pacific Ocean, from the Bering Sea to California

Pilot whale
Worldwide, except for polar waters

Polar bear
Arctic coastal waters and ice floes

Puma
Western regions of the U.S., except Alaska; all of South America, except Chile and part of Brazil

Raccoon
Southern Canada to Panama

Right whale (endangered)
North Atlantic Ocean, from Greenland to Florida

Sea otter
Pacific coast, along California, Canada, Alaska, and Russia

Snowshoe hare
Canada and Alaska, except arctic regions; Appalachian and Rocky Mountains in the U.S.

Walrus
Arctic seas and coastal areas

West Indian manatee (endangered)
Atlantic coast, from Florida to Virginia; coastal waterways along Gulf of Mexico and Caribbean Sea

White-tailed deer
Southern Canada to northern South America

Tables of Measures

Metric System

Units of Length	Units of Area
1 kilometer (km) = 1,000 meters (m)	1 square meter = 10,000 square centimeters
1 meter (m) = 10 decimeters (dm)	1 sq m = 10,000 sq cm
= 100 centimeters (cm)	
= 1,000 millimeters (mm)	1 square centimeter = 100 square millimeters
1 decimeter (dm) = 10 centimeters (cm)	1 sq cm = 100 sq mm
1 centimeter (cm) = 10 millimeters (mm)	

Units of Weight	Units of Volume
1 metric ton (t) = 1,000 kilograms (kg)	1 cubic meter = 1,000,000 cubic centimeters
1 kilogram (kg) = 1,000 grams (g)	1 cu m = 1,000,000 cu cm
1 gram (g) = 1,000 milligrams (mg)	1 cubic centimeter = 1,000 cubic millimeters
	1 cu cm = 1,000 cu mm

Units of Capacity	
1 kiloliter (kL) = 1,000 liters (L)	1 liter (L) = 1,000 milliliters (mL)

U.S. Customary System

Units of Length	Units of Area
1 mile (mi) = 1,760 yards (yd)	1 square yard (sq yd) = 9 square feet (sq ft)
= 5,280 feet (ft)	= 1,296 square inches (sq in.)
1 yard (yd) = 3 feet (ft)	1 square foot (sq ft) = 144 square inches (sq in.)
= 36 inches (in.)	
1 foot (ft) = 12 inches (in.)	

Units of Weight	Units of Volume
1 pound (lb) = 16 ounces (oz)	1 cubic yard (cu yd) = 27 cubic feet (cu ft)
1 ton (T) = 2,000 pounds (lb)	1 cubic foot (cu ft) = 1,728 cubic inches (cu in.)

Units of Capacity	
1 gallon (gal) = 4 quarts (qt)	1 cup (c) = 8 fluid ounces (fl oz)
1 quart (qt) = 2 pints (pt)	1 fluid ounce (fl oz) = 2 tablespoons (tbs)
1 pint (pt) = 2 cups (c)	1 tablespoon (tbs) = 3 teaspoons (tsp)

Units of Time

1 millennium = 10 centuries = 100 decades = 1,000 years (yr)	1 month (mo) = 28, 29, 30, or 31 days
1 century (cent.) = 10 decades = 100 years (yr)	1 week (wk) = 7 days
1 year (yr) = 12 months (mo) = 52 weeks (wk) plus 1 or 2 days = 365 or 366 days	1 day = 24 hours (hr)
	1 hour (hr) = 60 minutes (min)
	1 minute (min) = 60 seconds (sec)

Units of Body Measure

1 **digit** is the width of a finger.

1 **hand** is the width of the palm and thumb.

1 **span** is the distance from the tip of the thumb to the tip of the first (index) finger. Both fingers must be stretched as far apart as possible.

1 **cubit** is the length from the point of the elbow to the tip of the extended middle finger.

1 **yard** is the distance from the middle of the chest to the tip of the middle finger. The arm must be stretched out at the side of the body, and at a right angle with the body. 1 yard = $\frac{1}{2}$ fathom

1 **fathom** is the length of outstretched arms, measured across the chest from the tip of one middle finger to the tip of the other. 1 fathom = 2 yards

System Equivalents

1 inch is about 2.5 centimeters.

1 centimeter is about 0.4 inch.

1 kilometer is about 0.6 mile.

1 mile is about 1.6 kilometers.

1 meter is about 39 inches.

1 liter is about 1.1 quarts.

1 ounce is about 28 grams.

1 kilogram is about 2.2 pounds.

Roman Numerals

A **numeral** is a symbol used to represent a number. **Roman numerals,** developed about 500 B.C., use letters to represent numbers.

Roman Numeral	Number
I	1
V	5
X	10
L	50
C	100
D	500
M	1,000

Seven different letters are used in Roman numerals. Each letter stands for a different number.

A string of letters means that their values should be added together. For example, CCC = 100 + 100 + 100 = 300, and CLXII = 100 + 50 + 10 + 1 + 1 = 162.

If a smaller value is placed *before* a larger value, the smaller value is subtracted instead of added. For example, IV = 5 − 1 = 4, and CDX = 500 − 100 + 10 = 410.

The letters I (1), X (10), C (100), and M (1,000) are the only letters that may be subtracted. The letters V (5), L (50), and D (500) may not be subtracted. For example, 95 in Roman numerals is XCV. Writing VC for 95 is incorrect, because the letter V may not be subtracted.

The largest Roman numeral, M, stands for 1,000. One way to write large numbers is to write a string of Ms. For example, MMMM stands for 4,000. Another way to write large numbers is to write a bar above a numeral. The bar means that the numeral beneath should be multiplied by 1,000. So, $\overline{\text{IV}}$ also stands for 4,000. And $\overline{\text{M}}$ stands for 1,000 * 1,000 = 1 million.

Problem Solving

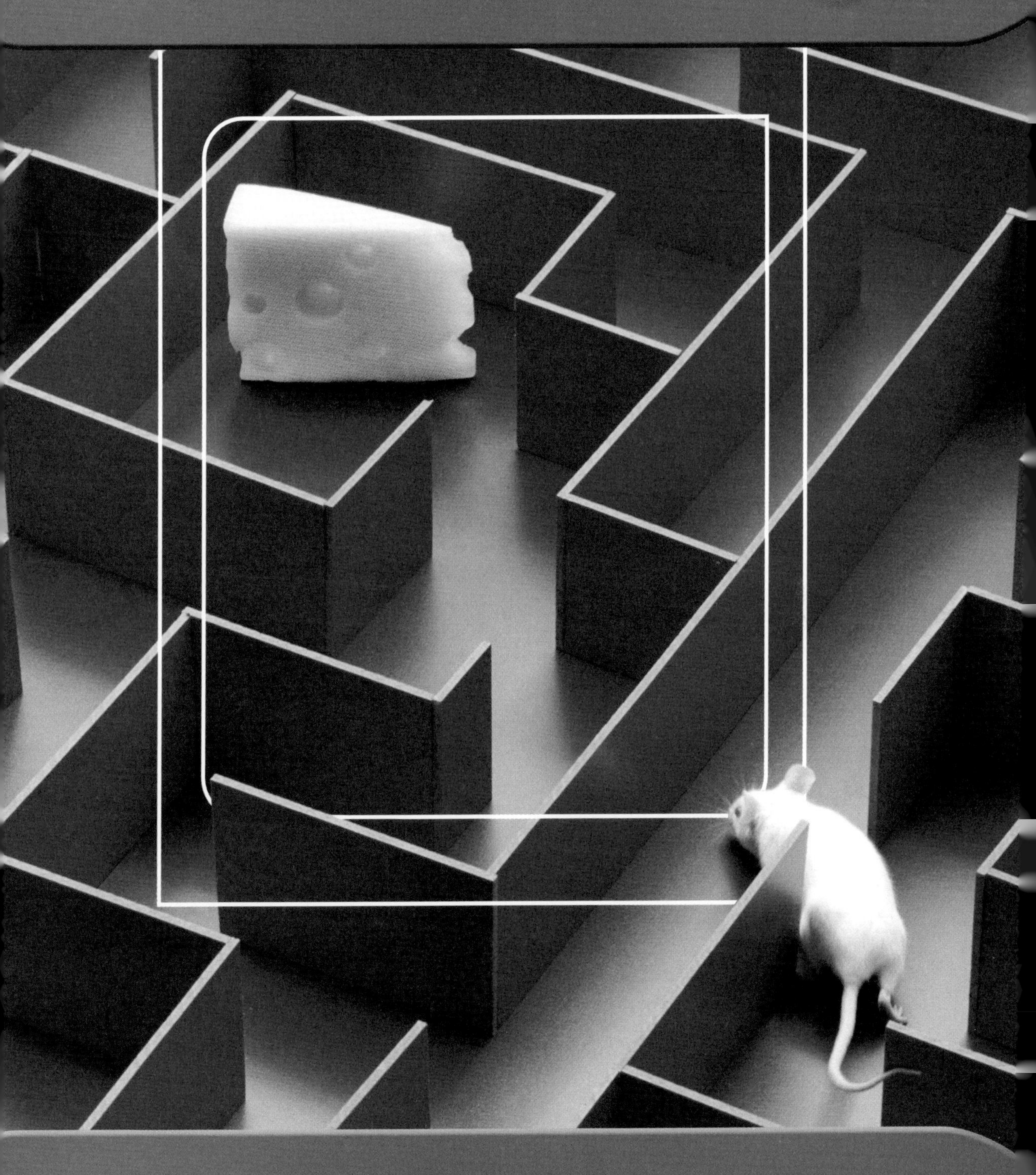

Solving Number Stories

Solving number stories is a big part of mathematics. Good problem solvers often follow a few simple steps every time they solve a number story. When you are trying to solve a number story, you can follow these same steps.

A Guide to Solving Number Stories

1. What do you understand from the story?
2. What will you do? Do it. Record what you did.
3. Answer the question. If you can, write a number model to show what you did.
4. Check. Ask, "Does my answer make sense? How do I know?"

These steps can take a lot of work.

1. What do you understand from the story?

- Read the story and retell it in your own words.
- What do you want to find out? Ask the question in your own words.
- Is the answer a number? Is the answer a length or other measurement?
- What do you know? Can you draw a picture or diagram to show what you know?

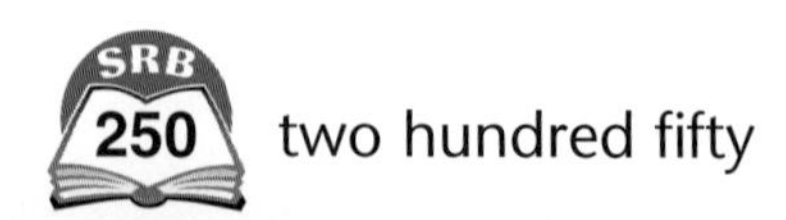

2. What will you do?

- Sometimes it's easy to know how to solve a problem.
- Other times you need to be creative.
- Is the problem like one that you have solved before?
- Is there a pattern that you can use?
- Can you compute to find the answer?
- Can you use counters, base-10 blocks, or some other tool?
- Can you make a table?
- Can you guess the answer and check to see if you're right?

Do it. Record what you did.

- Try to show how you solved the problem.
- Draw a picture.
- Write about what you did.

3. Answer the question.

- What are the units?
- Write a sentence that answers the question in the problem.
- If you can, write a number model that fits the problem.

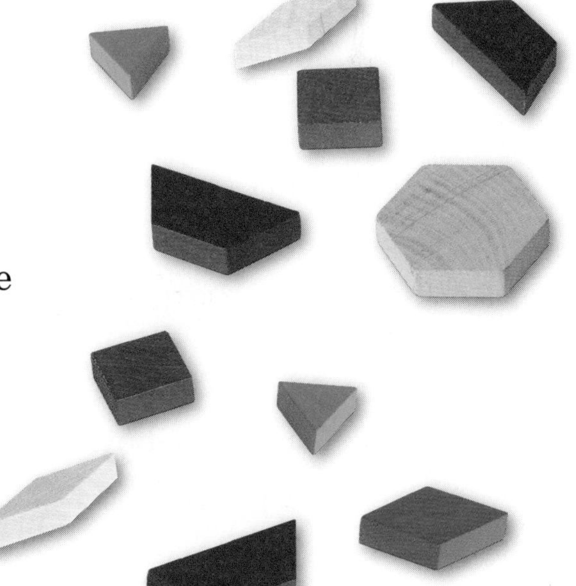

4. Check.

- Ask, "Does my answer make sense? How do I know?"
- Does your answer agree with other people's answers?
- Estimate the answer. Is your estimate close to your answer?

Check Your Understanding

1. Jennifer had 40 pennies. She put 25 pennies in her bank and gave the other pennies to three friends. How many pennies did each friend get if they got equal shares?
2. One side of a rectangle is 20 cm long. Another side is 5 cm long. What is the perimeter of the rectangle?
3. Rashid bought 4 markers for 56¢. How much did each marker cost?
4. The product of two numbers is 12. The sum of the numbers is 7. What are the numbers?
5. Mr. Cohen's third grade class has more boys than girls. If there were 2 more boys, there would be twice as many boys as girls. There are 8 girls in the class. How many boys are there?

Check your answers on page 343.

Guide to Solving Number Stories

1 **What do you understand from the story?**

- Read the story.
- What do you want to find out?
- What do you know?

2 **What will you do?**

- Draw a picture?
- Draw a diagram?
- Make tallies?
- Add?
- Subtract?
- Multiply?
- Divide?

Do it. Record what you did.

3 **Answer the question.**

Can you write a number model to show what you did?

4 **Check.**

Ask: "Does my answer make sense? How do I know?"

Change Number Stories

A number story is a **change story** if an amount is increased or decreased.

- If the amount is increased, we call it a **change-to-more** story.
- If the amount is decreased, we call it a **change-to-less** story.

Example 25 children are riding on a bus. Then 5 more children get on. How many children are on the bus now?

The number of children on the bus has increased.
This is a change-to-more number story.

A **change diagram** has spaces to show the **Start, Change,** and **End** numbers in a number story. It can help you solve a change number story.

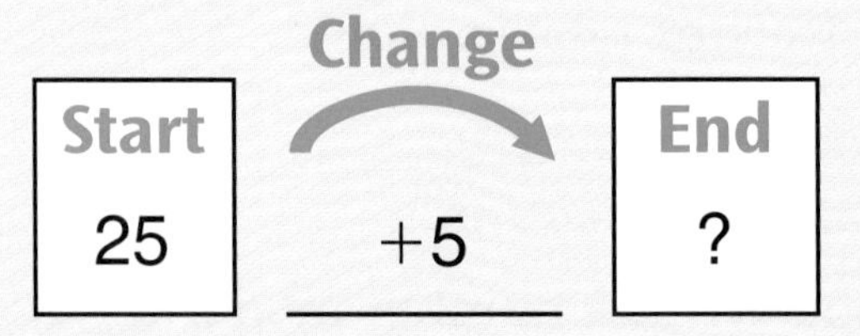

The change diagram shows the numbers you know and the number you need to find. Add 25 + 5 to solve the problem.

There are 30 children on the bus.

$25 + 5 = 30$ is a **number model** for this number story. The number model shows how the parts of the story are connected.

Example A bus leaves school with 42 children on it. At the first stop, 5 children get off. How many children are still on the bus?

Change

Start		End
42	−5	?

The change diagram shows the numbers you know and the number you need to find. You can subtract 42 − 5 to solve the problem.

There are 37 children left on the bus.

42 − 5 = 37 is a number model for this number story.

Example The temperature was 40 degrees at 7:00 A.M. By noon, it was 54 degrees. What was the temperature change?

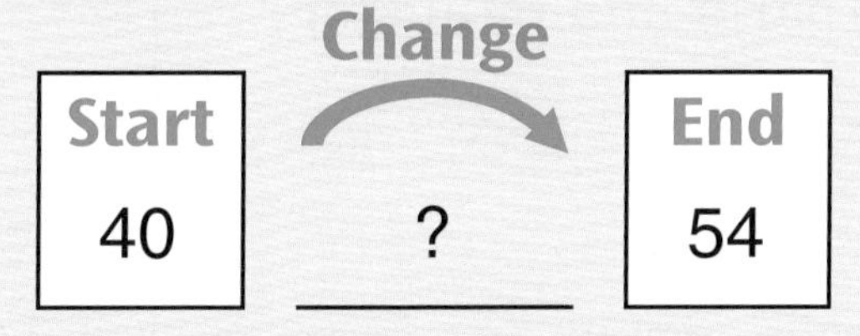

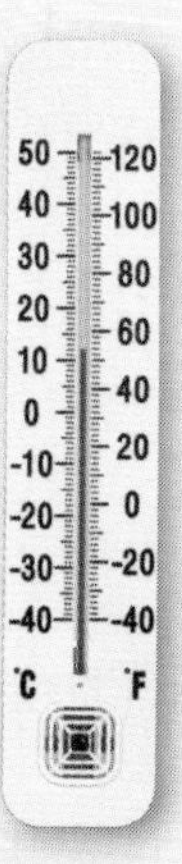

In some change stories, you know the Start and the End. You need to find the Change. Ask yourself, "What do I need to add to 40 to get 54?" The answer is 14.

The temperature increased 14 degrees.

40 + 14 = 54 is a number model for this number story.

Parts-and-Total Number Stories

A number story is a **parts-and-total story** if two or more parts are combined to form a total. Here are some simple examples of parts-and-total stories.

- Richard earned $6. Daniel earned $15. Together they earned $21. The two parts are $6 and $15. The total earnings is $21.
- A math quiz had 25 problems. Samantha had 21 correct answers and 4 incorrect answers. The total is 25 answers. The two parts are 21 answers and 4 answers.

Example There are 14 boys and 11 girls in Ms. Wilson's class. How many children are in her class?

A **parts-and-total diagram** has spaces to show each **Part** and the **Total** in a number story. It can help you solve a parts-and-total story.

The Parts are known.

You are looking for the Total.

Add 14 + 11 to solve the problem.

Total	
?	
Part	Part
14	11

There are 25 children in Ms. Wilson's class.

14 + 11 = 25 is a **number model** for this number story.

The number model shows how the different parts of the story are connected.

In some parts-and-total stories, you know the total but not all of the parts. You need to find one of the parts.

Example 35 children are riding on a bus. 20 of them are boys. How many girls are riding on the bus?

Total	
35	
Part	Part
20	?

The Total is known.

And one Part of the Total is known.

You are looking for the other Part.

One way to solve the problem is to ask yourself, "What do I need to add to 20 to get 35?" The answer is 15.

There are 15 girls on the bus.
$20 + 15 = 35$ is one number model for this number story.

Another way to solve the problem is to subtract the number of boys from the total number of children.

$35 - 20 = 15$, so there are 15 girls on the bus.
$35 - 20 = 15$ is another number model for this number story.

Check Your Understanding

Ulla spent 80 minutes reading a book and drawing a picture. She drew for 36 minutes. How long did she read?

Draw a parts-and-total diagram to help you solve the problem.

Check your answers on page 343.

Comparison Number Stories

In a **comparison story,** two quantities are compared. The **difference** between these quantities tells how much more or less one quantity is than the other.

Example There are 12 third graders and 8 second graders. How many more third graders are there than second graders?

A **comparison diagram** has spaces to show each **Quantity** and the **Difference** in a number story. It can help you solve a comparison number story.

The diagram shows that two Quantities are known.

You are looking for the Difference.

One way to solve the problem is to ask yourself, "What do I add to 8 to get 12?"

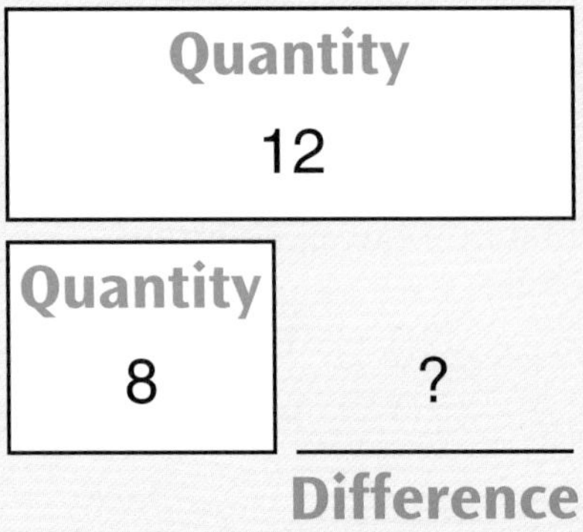

Quantity 12	
Quantity 8	? Difference

There are 4 more third graders.
$8 + 4 = 12$ is one **number model** for this number story.

Another way to solve the problem is to subtract the smaller number from the larger number. The difference is $12 - 8$.

The answer is 4.
Another number model for this number story is $12 - 8 = 4$.

Diagrams for Equal-Groups Problems

It often helps to fill in a diagram as you solve an equal-groups problem. The diagram has spaces to keep track of three things:

- the number of groups
- the number of objects in each group
- the total number of objects

Fill in the diagram with the numbers you know. Then write a question mark (?) for the number you want to find.

When do you multiply? If the total number of objects is not known, then you multiply to find it.

Example There are 3 rows with 5 chairs in each row. How many chairs are there in all?

There are 3 groups of 5.

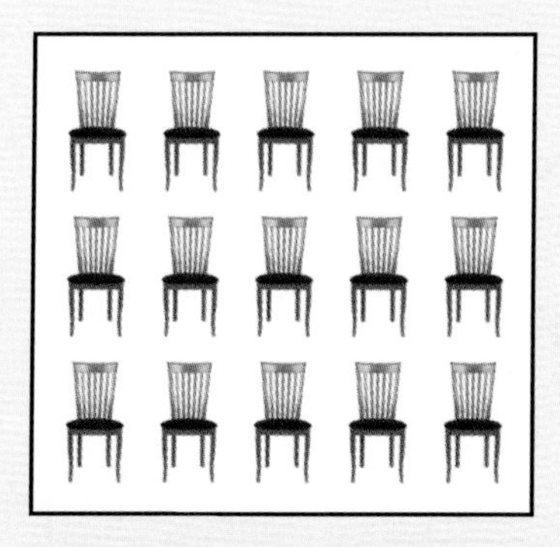

Groups	Objects in each group	Total objects
rows	chairs per row	chairs
3	5	?

To find the total number of chairs, multiply 3 by 5.

There are 15 chairs in all.
$3 \times 5 = 15$ is a number model for this problem.

When do you divide? If the total number of objects is known, then you divide to find the missing number.

Example 24 cards are placed in 4 equal piles. How many cards go in each pile?

You know the total number of objects and the number of piles (groups). You need to find the number of objects in each pile.

Divide the total number of cards by the number of piles. Divide 24 by 4.

Groups	Objects in each group	Total objects
piles	cards per pile	cards
4	?	24

There are 6 cards in each pile.

$24 \div 4 = 6$ is a number model for this problem.

Example Each table must have 6 chairs. There are 33 chairs. How many tables can have 6 chairs?

You know the total number of objects and the number of objects per group. You need to find the number of groups.

Divide the total number of chairs by the number of chairs in 1 group. Divide 33 by 6.

Groups	Objects in each group	Total objects
tables	chairs per table	chairs
?	6	33

Five tables can have 6 chairs. There are 3 chairs left over.

$33 \div 6 = 5$ (remainder 3) is a number model for this problem.

Calculators

About Calculators

Since kindergarten, you have used calculators to do mathematics. You use them to help you learn to count. You use them to operate with numbers.

Not all calculators are alike. Here is one type of calculator you may be using.

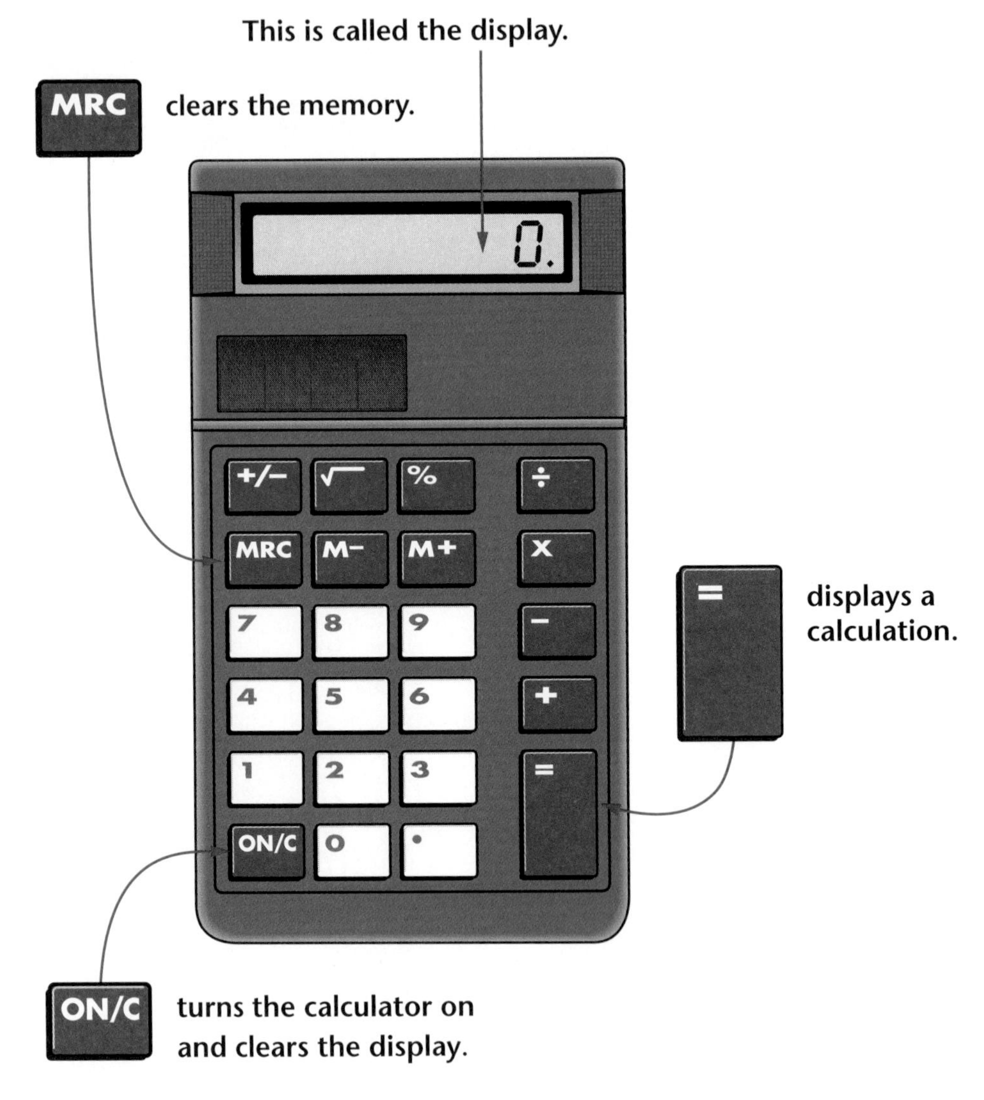

Here is another type of calculator you may be using.

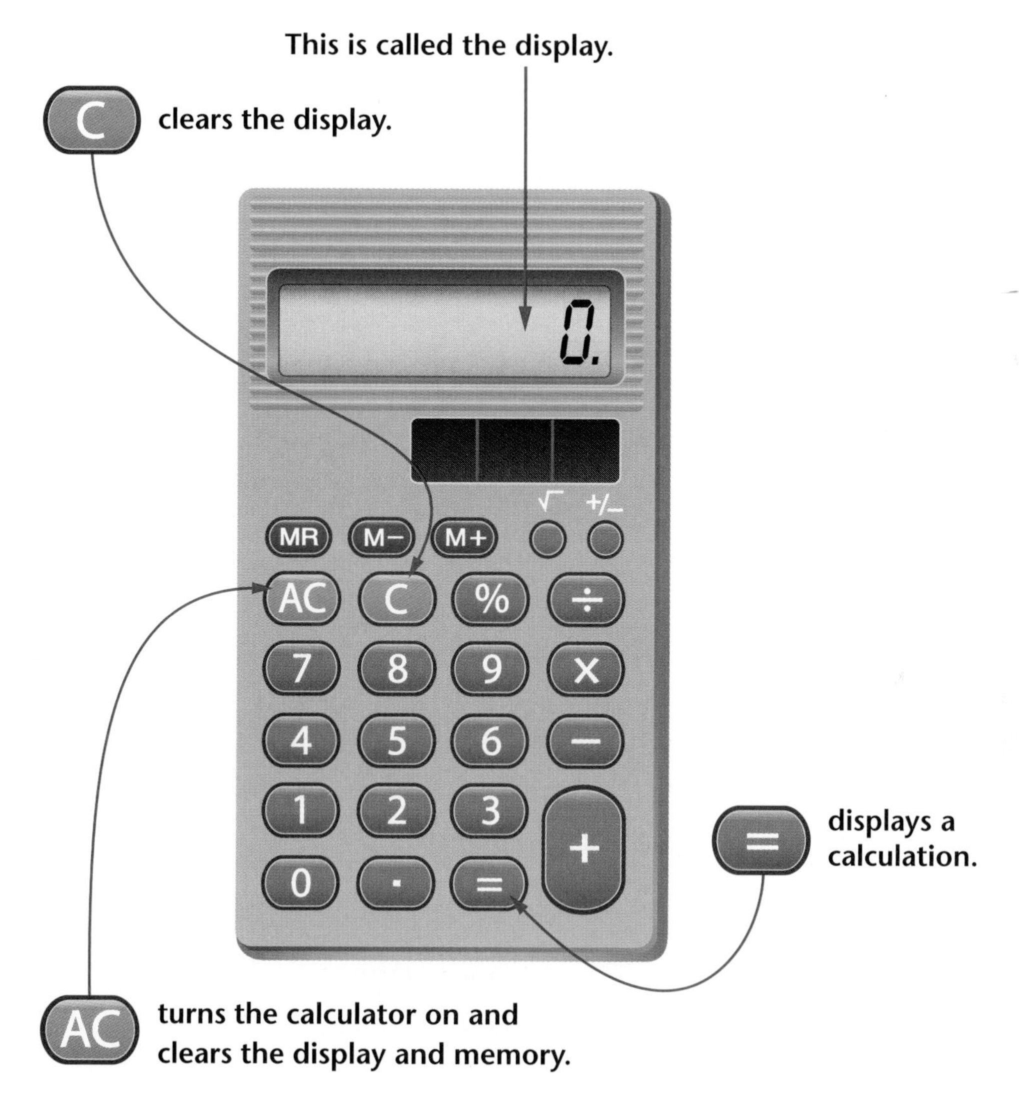

Take care of your calculator. Don't drop it, or leave it on a heater or in the sun.

Basic Operations

Note

Both Calculator A and Calculator B turn themselves off after a few minutes. So neither has an "Off" key.

Pressing a key is called **entering** instructions into the calculator. In this book and in your journal, most keys are shown in boxes such as [AC], [ON/C], and [=].

The simplest entry is to turn the calculator on. Another simple entry is to clear numbers from the display and memory. When it is cleared, 0. is in the display.

Calculator A	
Key	**Purpose**
[ON/C]	Turn the display on.
[ON/C] [ON/C]	Clear the display and memory.
[ON/C]	Clear only the display.

Calculator B	
Key	**Purpose**
(AC)	Turn the display on.
(AC)	Clear the display and memory.
(C)	Clear only the display.

The set of keys you press to do a calculation is called a **key sequence.** Key sequences to add, subtract, multiply, and divide numbers are shown below. The key sequences work for both the calculators shown. Remember to clear your calculator before starting a new problem.

Operation	Problem	Key sequence	Display
Add [+]	23 + 19	23 [+] 19 [=]	42.
Subtract [−]	42 − 19	42 [−] 19 [=]	23.
Multiply [X]	6 × 14	6 [X] 14 [=]	84.
Divide [÷]	84 ÷ 14	84 [÷] 14 [=]	6.

When you use a calculator, ask yourself if the number in the display makes sense. This helps you to decide if you made a mistake entering a number or operation.

Skip Counting on a Calculator

You can program your calculator to **skip count** up or back. The program you enter needs to tell the calculator four things:

- what number to count by
- whether to count up or down
- what number to start at
- when to count

The order of the steps in the program depends on your calculator. Here's how to program Calculator A.

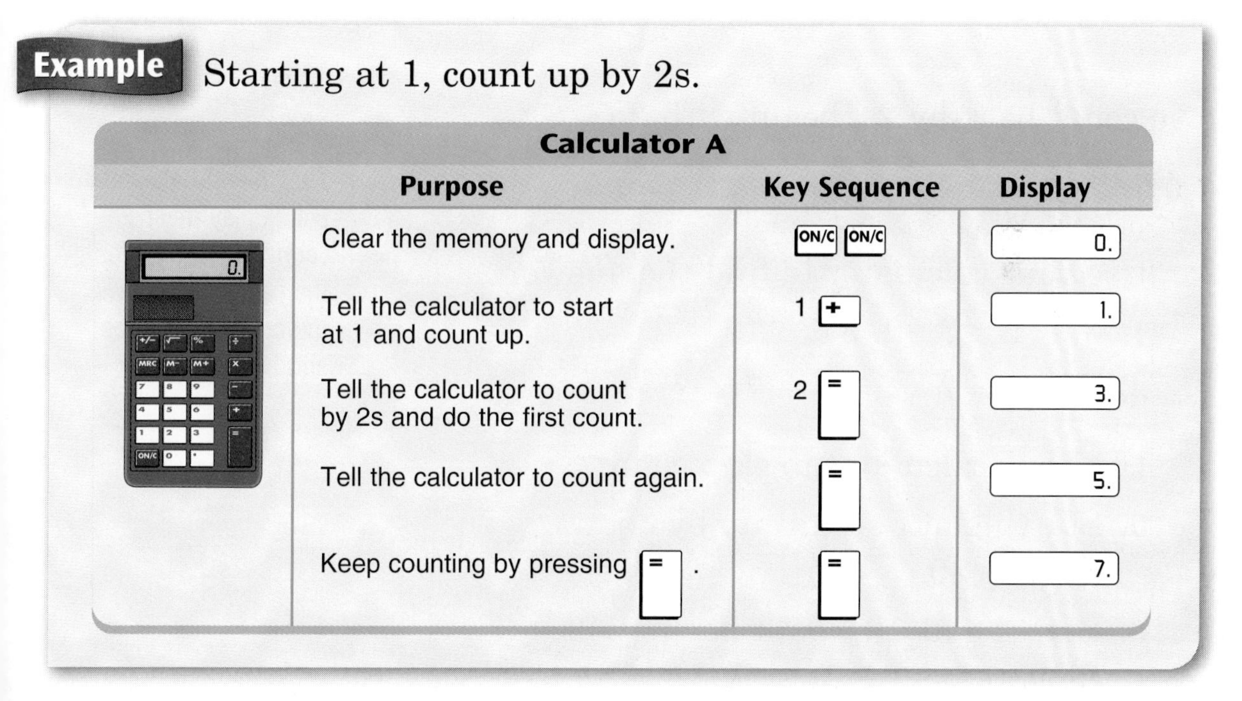

Example Starting at 1, count up by 2s.

Calculator A			
	Purpose	Key Sequence	Display
	Clear the memory and display.	ON/C ON/C	0.
	Tell the calculator to start at 1 and count up.	1 +	1.
	Tell the calculator to count by 2s and do the first count.	2 =	3.
	Tell the calculator to count again.	=	5.
	Keep counting by pressing =.	=	7.

To count back by 2s, begin with the starting number followed by −.

Here's how to program Calculator B.

Example Starting at 1, count up by 2s.

Calculator B			
Purpose		Key Sequence	Display
	Clear the memory and display.	(AC)	0.
	Tell the calculator to count up by 2.	2 (+) (+)	K 2.
	Tell the calculator to start at 1 and do the first count.	1 (=)	K 3.
	Tell the calculator to count again.	(=)	K 5.
	Keep counting by pressing (=).	(=)	K 7.

To count back by 2s, begin with 2 (−) (−).

Note

The "K" on Calculator B's display means "constant." It means the calculator knows the count by number and by direction.

Check Your Understanding

Use your calculator to find the answer.

1. $3 + 4 = ?$ **2.** $31 - 19 = ?$

3. $6 \times 5 = ?$ **4.** $56 \div 7 = ?$

Use your calculator to skip count.

5. Starting at 5, count up by 3s. Stop at 20.

6. Starting at 10, count back by 2s. Stop at 0.

Check your answers on page 343.

Games

Games

Throughout the year, you will play games that help you practice important math skills. Playing mathematics games gives you a chance to practice math skills in a way that is different and enjoyable. In this section of your *Student Reference Book,* you will find the directions for many games.

Materials

You need a deck of number cards for many of the games. You can use an Everything Math Deck, a deck of regular playing cards, or make your own deck out of index cards.

An Everything Math Deck includes 54 cards. There are 4 cards each for the numbers 0–10. And there is 1 card for each of the numbers 11–20.

You can also use a deck of regular playing cards after making a few changes. A deck of playing cards includes 54 cards (52 regular cards, plus 2 jokers). To create a deck of number cards, use a permanent marker to mark the cards in the following way:

- Mark each of the 4 aces with the number 1.
- Mark each of the 4 queens with the number 0.
- Mark the 4 jacks and 4 kings with the numbers 11 through 18.
- Mark the 2 jokers with the numbers 19 and 20.

For some games you will have to make a gameboard, a score sheet, or a set of cards that are not number cards. The instructions for doing this are included with the game directions. More complicated gameboards and card decks are available from your teacher.

Addition Top-It

Materials □ number cards 0–10 (4 of each)

Players 2 to 4

Skill Addition facts 0 to 10

Object of the game To collect the most cards.

Directions

1. Shuffle the cards. Place the deck number-side down on the table.
2. Each player turns over 2 cards and calls out the sum of the numbers.
3. The player with the largest sum wins the round and takes all the cards.
4. In case of a tie for the largest sum, each tied player turns over 2 more cards and calls out the sum of the numbers. The player with the largest sum then takes all the cards from both plays.
5. The game ends when not enough cards are left for each player to have another turn.
6. The player with the most cards wins.

Example Ann turns over a 6 and a 7. She calls out 13. Joe turns over a 10 and a 4. He calls out 14. Joe has the larger sum. He takes all 4 cards.

Angle Race

Materials ☐ 24-pin circular geoboard or a sheet of Circular-Geoboard Paper (*Math Masters,* p. 430)

☐ 15 rubber bands, or a straightedge and a pencil

☐ deck of *Angle Race* Degree-Measure Cards (*Math Masters,* p. 441)

Players 2

Skill Recognizing angle measures

Object of the game To complete an angle exactly at the 360° mark on a circular geoboard.

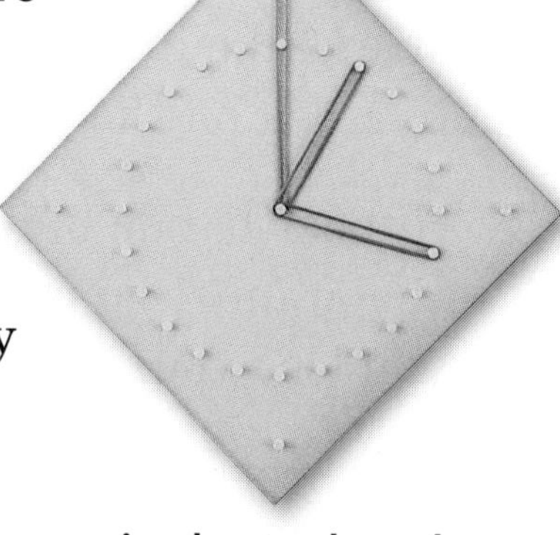

circular geoboard

Directions

1. Shuffle the cards. Place the deck number-side down on the table.
2. If you have a circular geoboard, stretch a rubber band from the center peg to the 0° peg.

   If you do *not* have a circular geoboard, use a sheet of Circular-Geoboard Paper. Draw a line segment from the center dot to the 0° dot. Instead of stretching rubber bands, you will draw line segments.

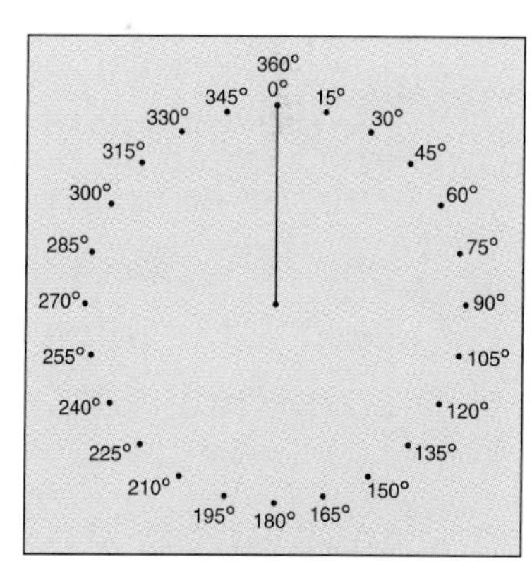

Circular-Geoboard Paper

3. Players take turns. Both players use the same geoboard (or Circular-Geoboard Paper).

4. When it is your turn, select the top degree-measure card. Make an angle on the geoboard that has the same degree measure as shown on the card. Use the last rubber band placed on the geoboard as one side of your angle. Make the second side of your angle by stretching another rubber band from the center peg to a peg on the circle, going *clockwise*.

5. Rubber bands may not go past the 360° (or 0°) peg. If you must go past the 360° peg to make an angle, you lose your turn.

6. The first player to complete an angle exactly on the 360° peg wins.

Example The first player draws a 30° card. The player makes a 30° angle by stretching a rubber band from the center peg to the 30° peg. The second player draws a 75° card. This player makes a 75° angle by stretching a rubber band from the center peg to the 105° peg. Players continue to take turns, stretching rubber bands clockwise around the circle, until one player exactly reaches the 360° peg.

Array Bingo

Materials ☐ 1 set of *Array Bingo* Cards for each player (*Math Masters,* p. 442)
☐ number cards 1–20 (1 of each)

Players 2 or 3

Skill Multiplication for arrays and equal groups

3×3

2×3

Object of the game To have a row, column, or diagonal of cards facedown.

Directions

1. Each player arranges his or her array cards faceup in a 4-by-4 array.
2. Shuffle the number cards. Place them number-side down.
3. Players take turns. When it is your turn, draw a number card. Look for any one of your array cards with that number of dots and turn it facedown. If there is no matching array card, your turn ends. Place your number card in a discard pile.
4. The first player to turn a card facedown so that a row, column, or diagonal of cards is all facedown, calls out "Bingo!"
5. If all the number cards are used before someone wins, shuffle the deck and continue playing.

Example Mary draws the number card 4.

She turns over the card with the 2×2 array and calls out "Bingo."

4×3

3×5

2×3

2×2

3×3

Baseball Multiplication

Materials ☐ 1 *Baseball Multiplication* game mat (*Math Masters,* p. 443)
☐ 2 six-sided dice
☐ 4 counters

Players 2 teams of one or more players each

Skill Multiplication facts 1 to 6

Object of the game To score more runs in a 3-inning game.

Directions

The rules are similar to the rules for baseball, but this game lasts only 3 innings. In each inning, each team bats until it makes 3 outs. Teams flip a coin to decide who bats first. The team with more runs when the game is over wins.

Pitching and batting: Members of the team not at bat take turns "pitching." They roll the two dice to get 2 factors. Players on the "batting" team take turns multiplying the 2 factors and saying the product.

The pitching team checks the product. (Use a calculator or the Multiplication/Division Facts Table on page 52.) An incorrect answer is a strike, and another pitch (dice roll) is thrown. Three strikes make an out.

Hits and runs: If the answer is correct, the batter checks the Scoring Chart on the game mat. If the chart shows a hit, the batter moves a counter to a base as shown in the Scoring Chart. Runners already on base are moved ahead of the batter by the same number of bases. A run is scored every time a runner crosses home plate.

Keeping score: For each inning, keep a tally of runs scored and outs made. Use the Runs and Outs Tally on the game mat. At the end of the inning, record the number of runs on the Scoreboard.

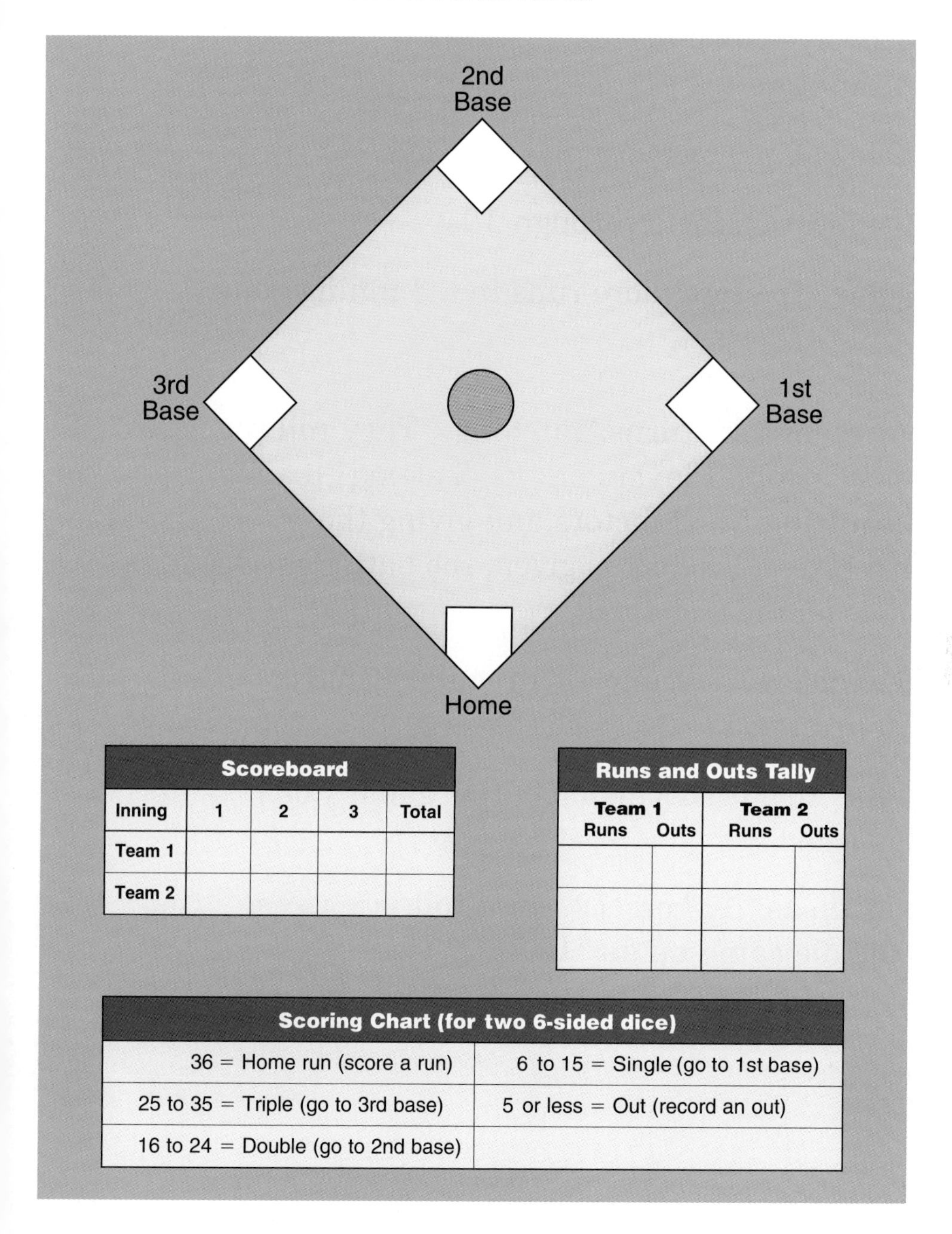

Scoreboard				
Inning	1	2	3	Total
Team 1				
Team 2				

Runs and Outs Tally			
Team 1		Team 2	
Runs	Outs	Runs	Outs

Scoring Chart (for two 6-sided dice)	
36 = Home run (score a run)	6 to 15 = Single (go to 1st base)
25 to 35 = Triple (go to 3rd base)	5 or less = Out (record an out)
16 to 24 = Double (go to 2nd base)	

Baseball Multiplication (Advanced Version)

Materials ☐ 1 *Baseball Multiplication* (Advanced) game mat (*Math Masters,* p. 444)
☐ 1 twelve-sided die
☐ 4 counters

Players 2 teams of one or more players each

Skill Multiplication facts through 12s

Object of the game To score more runs in a 3-inning game.

Directions

Members of one team take turns "pitching." They roll the die twice to get 2 factors. Players on the "batting" team take turns multiplying the 2 factors and giving the product. When a correct product is given, the batter checks the Scoring Chart on the game mat.

The rest of the game is the same as a regular game of *Baseball Multiplication.*

You can make the Advanced Version of this game a bit easier with this rule:

If the die comes up as "11" or "12" on a roll, pretend that the die came up as "10."

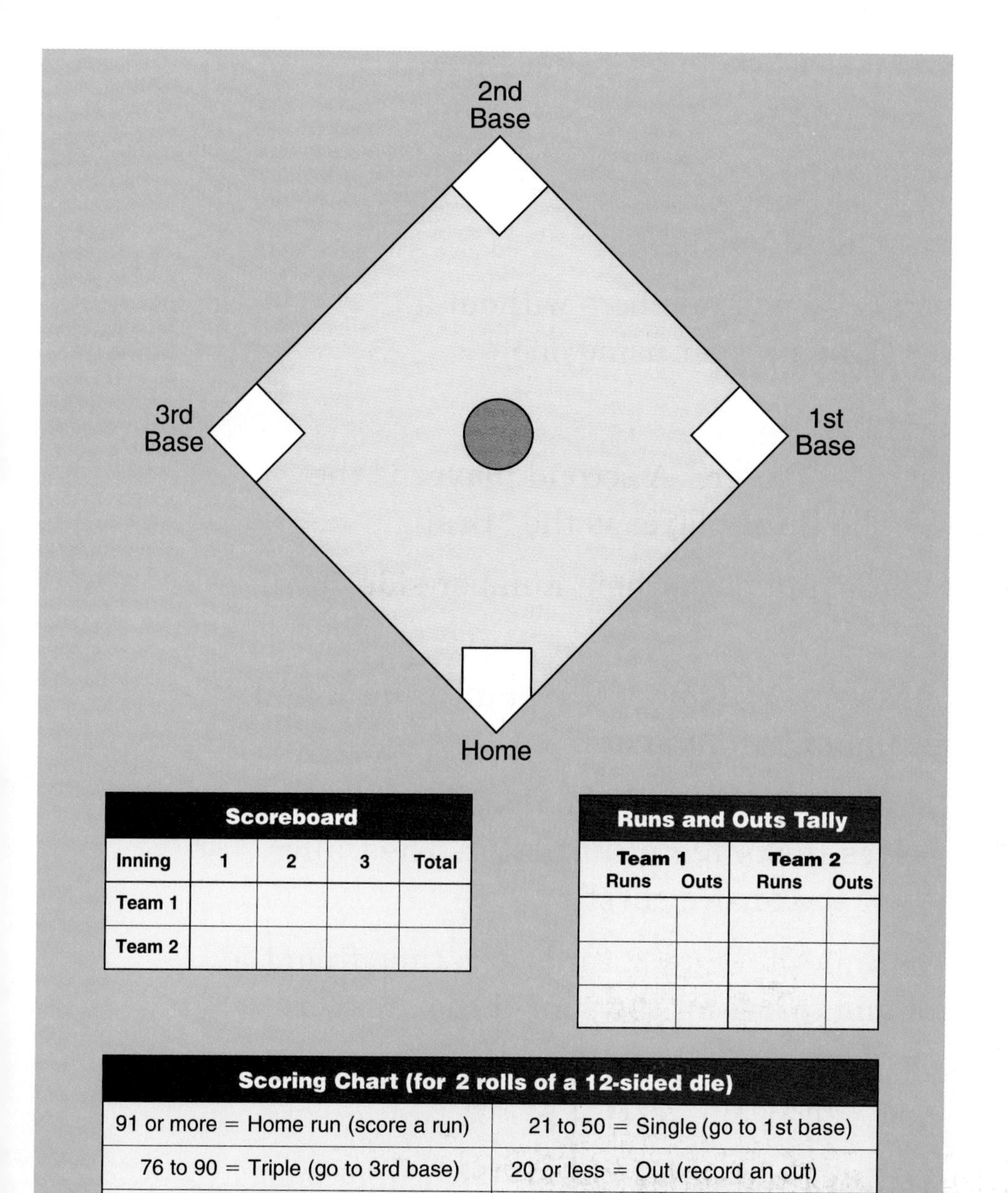

Scoreboard				
Inning	**1**	**2**	**3**	**Total**
Team 1				
Team 2				

Runs and Outs Tally			
Team 1		**Team 2**	
Runs	**Outs**	**Runs**	**Outs**

Scoring Chart (for 2 rolls of a 12-sided die)	
91 or more = Home run (score a run)	21 to 50 = Single (go to 1st base)
76 to 90 = Triple (go to 3rd base)	20 or less = Out (record an out)
51 to 75 = Double (go to 2nd base)	

Beat the Calculator (Addition)

Materials ☐ number cards 0–9 (4 of each)
☐ 1 calculator

Players 3

Skill Mental addition skills

Object of the game To add numbers without a calculator faster than a player using one.

Directions

1. One player is the "Caller." A second player is the "Calculator." The third player is the "Brain."
2. Shuffle the cards and place them number-side down on the table.
3. The Caller draws 2 cards from the number deck and asks for the sum of the numbers.
4. The Calculator solves the problem *with* a calculator. The Brain solves it *without* a calculator. The Caller decides who got the answer first.
5. The caller continues to draw 2 cards at a time from the number deck and to ask for the sum of the numbers.
6. Players trade roles every 10 turns or so.

Example The Caller draws a 7 and a 9. The Caller says, "7 plus 9." The Brain and the Calculator each solve the problem. The Caller decides who got the answer first.

Beat the Calculator (Multiplication)

Materials ☐ number cards 1–10 (4 of each)
☐ 1 calculator

Players 3

Skill Mental multiplication skills

Object of the game To multiply numbers without a calculator faster than a player using one.

Directions

1. One player is the "Caller." A second player is the "Calculator." The third player is the "Brain."
2. Shuffle the cards and place them number-side down on the table.
3. The Caller draws 2 cards from the number deck and asks for the product of the numbers.
4. The Calculator solves the problem *with* a calculator. The Brain solves it *without* a calculator. The Caller decides who got the answer first.
5. The Caller continues to draw 2 cards at a time from the number deck and ask for the product of the numbers.
6. Players trade roles every 10 turns or so.

Example The Caller draws a 10 and a 7. The Caller says, "10 times 7." The Brain and the Calculator each solve the problem. The Caller decides who got the answer first.

The Block-Drawing Game

Materials ☐ 1 paper bag

☐ 7 blocks (all the same size and shape) in 2 or 3 different colors

Players 3 or more

Skill Using chance data to estimate

Object of the game To correctly guess how many blocks of each color are in a bag.

Directions

1. Choose one player to be the "Director."
2. The Director secretly puts 3, 4, or 5 blocks (not all the same color) into a paper bag. The Director tells the other players *how many blocks* are in the bag, *but not their colors.*
3. Players take turns taking 1 block out of the bag, showing it, and replacing it.
4. After each draw, the Director records the color and keeps a tally on a slate or piece of paper.
5. A player may try to guess the colors of the blocks and the number of blocks of each color at any time.
6. If a player guesses incorrectly, that player is out of the game.
7. The first player to guess correctly wins the game.

Example The Director tells the players that there are 5 blocks in the bag.

green //
red //
blue /

tally after 5 draws

After 5 draws, Player 1 guesses 2 green, 2 red, and 1 blue. This guess is incorrect. Player 1 is out of the game.

green ///
red //
blue //

tally after 7 draws

After 7 draws, Player 2 guesses 2 green, 1 red, and 2 blue. This guess is incorrect. Player 2 is out of the game.

Player 3 then guesses 3 green, 1 red, and 1 blue. This guess is correct, and Player 3 wins the game.

Division Arrays

Materials ☐ number cards 6–18 (1 of each)
☐ 1 six-sided die
☐ 18 counters

Players 2 to 4

Skill Division (with remainder) for equal grouping

Object of the game To have the highest total score.

Directions

1. Shuffle the cards. Place the deck number-side down on the table.
2. Players take turns. When it is your turn, draw a card and take the number of counters shown on the card. You will use the counters to make an array.
   - Now roll the die. The number on the die is the number of equal rows you must have in your array.
   - Make an array with the counters.
   - Your score is the number of counters in one row. If there are no leftover counters, your score is double the number of counters in one row.
3. Players keep track of their scores. The player with the highest total score at the end of 5 rounds wins.

Example

Number card	Die	Array formed	Leftovers?	Score
10	2	● ● ● ● ● ● ● ● ● ●	no	10
9	2	● ● ● ● ● ● ● ●	● yes	4
14	3	● ● ● ● ● ● ● ● ● ● ● ●	●● yes	4

Equivalent Fractions Game

Materials ☐ 1 deck of Fraction Cards (*Math Journal* 2, Activity Sheets 5–8)

Players 2

Skill Recognizing fractions that are equivalent

Object of the game To collect more Fraction Cards.

Directions

1. Shuffle the Fraction Cards and place the deck picture-side down on the table.
2. Turn the top card over near the deck of cards.
3. Players take turns. When it is your turn, turn over the top card from the deck. Try to match this card with a picture-side up card on the table.
   - If you find a match, take the 2 matching cards. Then, if there are no cards left picture-side up, turn the top card over near the deck.
   - If you cannot find a match, place your card picture-side up next to the other cards. Your turn is over.
4. The game ends when all cards have been matched. The player with more cards wins.

Example The top card is turned over and put on the table. The picture shows $\frac{4}{6}$.

Player 1 turns over the $\frac{2}{3}$ card. This card matches $\frac{4}{6}$. Player 1 takes both cards. There are no cards left picture-side up. So Player 1 turns over the top card and puts it near the deck. The picture shows $\frac{6}{8}$.

Player 2 turns over the $\frac{0}{4}$ card. There is no match. This card is placed next to $\frac{6}{8}$. It is Player 1's turn again.

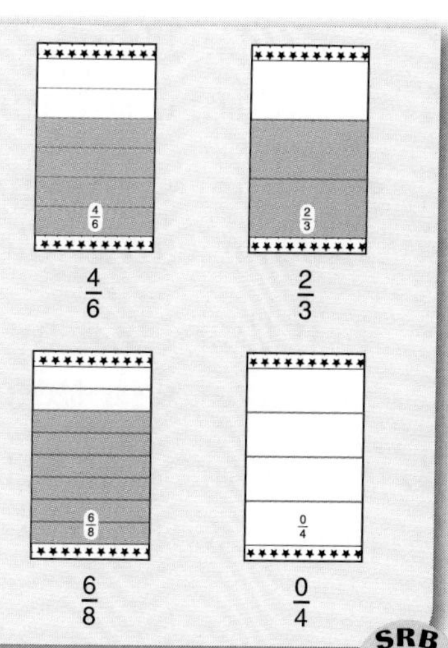

Equivalent Fractions Game
(Advanced Version)

Materials ☐ 1 deck of Fraction Cards (*Math Journal 2*, Activity Sheets 5–8)

Players 2

Skill Recognizing fractions that are equivalent

Object of the game To collect more Fraction Cards.

Directions

1. Shuffle the Fraction Cards and place the deck picture-side down on the table.
2. Turn the top card over near the deck of cards.
3. Players take turns. When it is your turn, take the top card from the deck, **but do not turn it over** (keep the picture side down). Try to match the fraction with one of the picture-side up cards on the table.
   - ◆ If you find a match, turn the card over to see if you matched the cards correctly. If you did, take both cards. Then, if there are no cards left picture-side up, turn the top card over.
   - ◆ If there is no match, place your card next to the other cards, picture-side up. Your turn is over.
   - ◆ If there is a match but you did not find it, the other player can take the matching cards.
4. The game ends when all cards have been matched. The player with more cards wins.

Factor Bingo

Materials ☐ number cards 2–9 (4 of each)
☐ 1 *Factor Bingo* game mat for each player (*Math Masters,* p. 448)
☐ 12 counters for each player

Players 2 to 4

Skill Finding factors of a number

Object of the game To get 5 counters in a row, column, or diagonal; or to get 12 counters anywhere on the game mat.

Directions

1. Fill in your own game mat. Choose 25 different numbers from the numbers 2 through 90.
2. Write each number you choose in exactly 1 square on your game mat grid. Be sure to mix the numbers up as you write them on the grid; they should not all be in order. To help you keep track of the numbers you use, circle them in the list below the game mat.
3. Shuffle the number cards and place them number-side down on the table. Any player can turn over the top card. This top card is the "factor."
4. Players check their grids for a number that has the card number as a factor. Players who find such a number cover the number with a counter. A player may place only 1 counter on the grid for each card that is turned over.
5. Turn over the next top card and continue in the same way. You call out "Bingo!" and win the game if you are the first player to get 5 counters in a row, column, or diagonal. You also win if you get 12 counters anywhere on the game mat.
6. If all the cards are used before someone wins, shuffle the cards again and continue playing.

Example A 5-card is turned over. So the number 5 is the "factor." Any player may place one counter on a number for which 5 is a factor, such as 5, 10, 15, 20, or 25. A player may place only one counter on the game mat for each card that is turned over.

Sample *Factor Bingo* Game Mat

Choose any 25 *different* numbers from the numbers 2 through 90. Write each number you choose in exactly 1 square on your game mat page. To help you keep track of the numbers you use, circle them in the list on your game mat page.

	2	3	4	5	6	7	8	9	10
11	12	13	14	15	16	17	18	19	20
21	22	23	24	25	26	27	28	29	30
31	32	33	34	35	36	37	38	39	40
41	42	43	44	45	46	47	48	49	50
51	52	53	54	55	56	57	58	59	60
61	62	63	64	65	66	67	68	69	70
71	72	73	74	75	76	77	78	79	80
81	82	83	84	85	86	87	88	89	90

Factor Bingo Game Mat

Fraction Top-It

Materials ☐ 1 deck of Fraction Cards (*Math Journal 2,* Activity Sheets 5–8)

Players 2

Skill Comparing fractions

Object of the game To collect more cards.

Directions

1. Shuffle the Fraction Cards and place the deck picture-side down on the table.
2. Each player turns over a card from the top of the deck. Players compare the shaded parts of the cards. The player with the larger fraction shaded takes both cards.
3. If the shaded parts are equal, the fractions are equivalent. Each player then turns over another card. The player with the larger fraction shaded takes all the cards from both plays.
4. The game is over when all cards have been taken from the deck. The player with more cards wins.

Examples Players turn over a $\frac{3}{4}$ card and a $\frac{4}{6}$ card.

The $\frac{3}{4}$ card has a larger shaded area. The player holding the $\frac{3}{4}$ card takes both cards.

Players turn over a $\frac{1}{2}$ card and a $\frac{4}{8}$ card.

The shaded parts are equal. Each player turns over another card. The player with the larger Fraction Card takes all the cards.

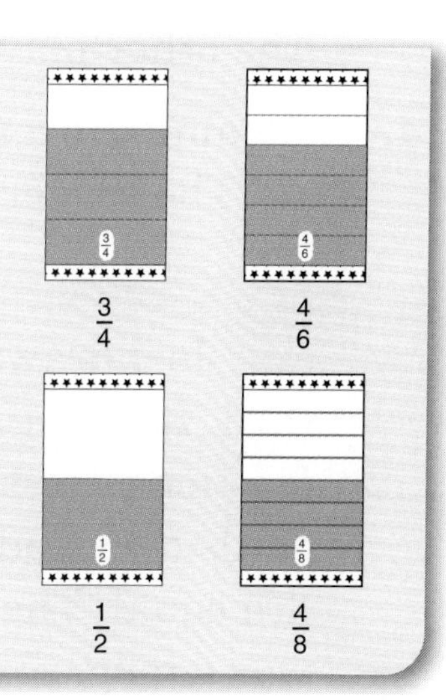

Fraction Top-It (Advanced Version)

Materials ☐ 1 deck of Fraction Cards (*Math Journal 2,* Activity Sheets 5–8)

Players 2

Skill Comparing fractions

Object of the game To collect more cards.

Directions

1. Shuffle the Fraction Cards and place the deck picture-side down on the table.
2. Each player takes a card from the top of the deck **but does not turn it over.** The cards remain picture-side down.
3. Players take turns. When it is your turn:
   - Say whether you think your fraction is greater than, less than, or equivalent to the other player's fraction.
   - Turn the cards over and compare the shaded parts. If you were correct, take both cards. If you were wrong, the other player takes both cards.
4. The game is over when all cards have been taken from the deck. The player with more cards wins.

Example Joel draws a $\frac{2}{8}$ card. Sue draws a $\frac{1}{4}$ card. It is Sue's turn, and she says that her fraction is less than Joel's. They turn their cards over and find that the shaded areas are equal. The fractions are equivalent. Sue was wrong, so Joel takes both cards.

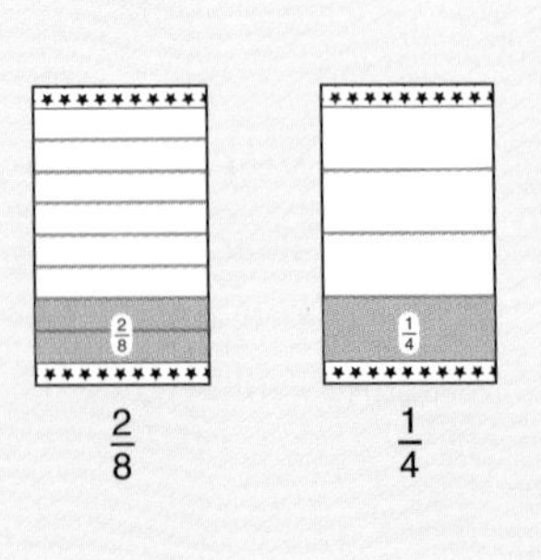

Less Than You!

Materials ☐ number cards 0–10 (4 of each)

Players 2

Skill Mental addition skills; developing a winning game strategy

Object of the game To say "Less than you!" and to have a sum that is less than the other player's.

Directions

Shuffle the cards. Deal 2 cards to each player, number-side down. Place the remaining deck of cards number-side down on the table. Players take turns. When it is your turn:

1. Take the top card from the deck. You now have 3 cards in your hand.
2. Discard the card in your hand with the largest number. Place this card *faceup* in a discard pile. (Discard means "take out of your hand and put aside.")
3. Add the 2 numbers on the cards left in your hand.
4. If you think that your sum is less than the other player's sum, say "Less than you!" If your sum *is* less, you win. If your sum is *not* less, you lose. The game is over.
5. If you don't say "Less than you!," your turn is over. The game is not over until one of the players says "Less than you!"

Advanced Version

Deal 3 cards to each player instead of 2.

Memory Addition/Subtraction

Materials ☐ 1 calculator

Players 2

Skill Mental addition and subtraction skills; using a calculator's memory keys

Object of the game To make the number in the memory of a calculator match a target number.

Directions

1. Players agree on a target number less than 50.
2. Either player clears the calculator's memory. (See **Using the Memory Keys** on the next page.) Both players must be able to see the calculator at all times.
3. Players take turns adding 1, 2, 3, 4, or 5 to the calculator's memory using the (M+) key, or subtracting 1, 2, 3, 4, or 5 from the memory using the (M-) key. They keep track of the results in their heads. A player cannot use the number that was just used by the other player.
4. The goal is to make the number in memory match the target number. When it is a player's turn and he or she thinks the number in memory is the same as the target number, the player says "same." Then he or she presses (MR) or [MRC] to display the number in memory.

   A player can say "same" and press (MR) or [MRC] before or after adding or subtracting a number.
5. If the number in the display matches the target number, the player who said "same" wins. If the number does not match the target number, the other player wins.

Using the Memory Keys

- Press (AC) or [MRC] [MRC] to clear the memory.
- Press (M+) to add the number in the display to memory.
- Press (M−) to subtract the number in the display from memory.
- Press (MR) or [MRC] once to display the number in memory.
- Change the directions if your calculator works differently.

Example Target number: 19

Winnie presses	Display shows	Maria presses	Display shows
4 (M+)	M 4	5 (M+)	M 5
3 (M+)	M 3	1 (M+)	M 1
2 (M−)	M 2	3 (M+)	M 3
5 (M+) (MR) or 5 (M+) [MRC]	M 19		

Winnie says "same" after pressing 5 (M+). Then she presses (MR) or [MRC] and the display shows the target number 19. Winnie wins. Either player presses (AC) or [MRC] [MRC] to clear the memory before starting a new game. Then, if the display shows a number that is different from 0, press [ON/C] to clear the display.

Missing Terms

Materials ☐ 1 calculator for each player

Players 2

Skill Mental addition and subtraction skills

Object of the game To say how one number was changed to obtain a second number.

Directions

1. Players enter the same number into their calculators.
2. One player secretly changes this number by adding or subtracting a number.
3. The other player is shown the new number that appears in the calculator display. He or she guesses what was done to the original number to get the new number.

Example Both calculators are set to 7.

Joyce secretly changes the display by pressing (+) 9 (=). The display now shows the number 16.

Joyce shows the display number 16 to Al.
Al says, "You added 9." He is correct.

Multiplication Bingo (Easy Facts)

Materials
- ☐ number cards 1–6 and 10 (4 of each)
- ☐ 1 *Multiplication Bingo* game mat for each player (*Math Masters,* p. 449)
- ☐ 8 counters for each player

Players 2 or 3

Skill Mental multiplication skills

Object of the game To get 4 counters in a row, column, or diagonal; or 8 counters anywhere on the game mat.

Directions

1. The game mat is shown below. You can make your own game mat on a piece of paper. Write each of the numbers in the list in one of the squares on the grid. Don't write the numbers in order. Mix them up.

List of Numbers	
1	18
4	20
6	24
8	25
9	30
12	36
15	50
16	100

***Multiplication Bingo* Game Mat**

2. Shuffle the number cards and place the deck number-side down on the table.

3. Players take turns. When it is your turn, take the top 2 cards and call out the product of the 2 numbers. If someone does not agree with your answer, check it by using the Multiplication/Division Facts Table on page 52 in your *Student Reference Book* or the inside front cover of your journal.

   - If your answer is incorrect, you lose your turn.
   - If your answer is correct and the product is a number on your game mat, place a counter on that number. You may only place a counter on your game mat when it is your turn.

4. If you are the first player to get 4 counters in a row, column, or diagonal, call out "Bingo!" and win the game! You can also call "Bingo!" and win if you get 8 counters anywhere on your game mat.

If all the cards are used before someone wins, shuffle the cards again and keep playing.

Example A player could call out "Bingo!" with any of these game mats:

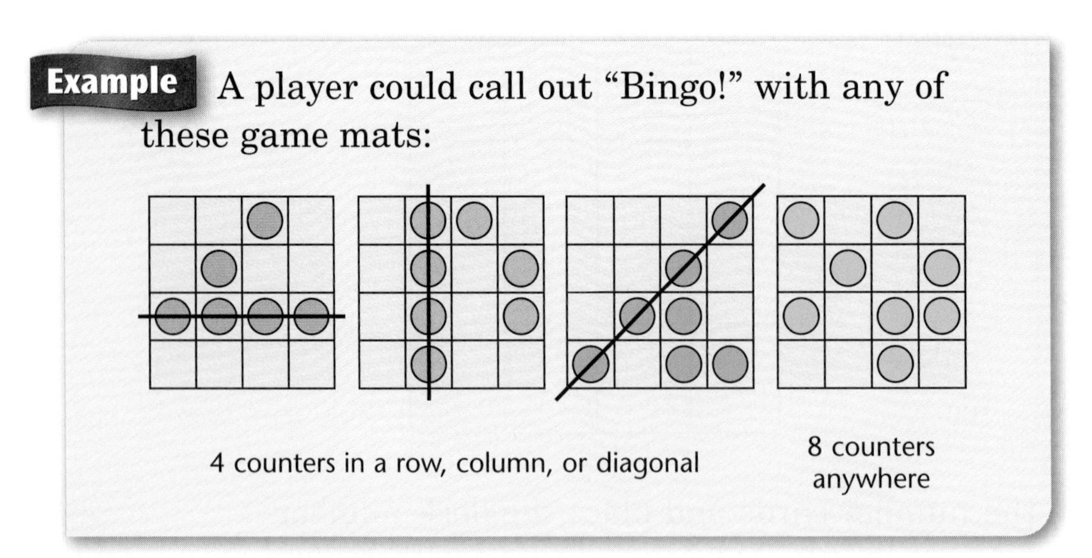

Multiplication Bingo (All Facts)

Materials ☐ number cards 2–9 (4 of each)
☐ 1 *Multiplication Bingo* Game Mat for each player (*Math Masters,* p. 449)
☐ 8 counters for each player

Players 2 or 3

Skill Mental multiplication skills

Object of the game To get 4 counters in a row, column, or diagonal; or 8 counters anywhere on the game mat.

Directions

1. The game mat is shown below. You can make your own game mat on a piece of paper. Write each of the numbers in the list in one of the squares on the grid. Don't write the numbers in order. Mix them up.

List of Numbers	
24	48
27	49
28	54
32	56
35	63
36	64
42	72
45	81

***Multiplication Bingo* Game Mat**

2. Follow the directions for playing *Multiplication Bingo* (Easy Facts).

Multiplication Draw

Materials □ number cards 1–5 and 10 (4 of each)
□ 1 *Multiplication Draw* Record Sheet (*Math Masters,* p. 450)

Players 2 or 3

Skill Multiplication facts

Object of the game To have the largest sum.

Directions

1. Shuffle the cards and place the deck number-side down.
2. Players take turns. When it is your turn, draw 2 cards from the deck to get 2 multiplication factors. Record both factors and their product on your Record Sheet.
3. After 5 turns, each player finds the sum of their 5 products.
4. The player with the largest sum wins the round.

Advanced Version

Include cards with numbers 6–9 in the number deck.

Example Alex draws a 3 card and a 10 card. He records $3 \times 10 = 30$ on his Record Sheet.

Alex	Round 1	Round 2	Round 3
1st draw:	3 × 10 = 30	__ × __ = __	__ × __ = __
2nd draw:	__ × __ = __	__ × __ = __	__ × __ = __
3rd draw:	__ × __ = __	__ × __ = __	__ × __ = __
4th draw:	__ × __ = __	__ × __ = __	__ × __ = __
5th draw:	__ × __ = __	__ × __ = __	__ × __ = __
Sum of products:	____	____	____

Multiplication Top-It

Materials □ number cards 0–10 (4 of each)

Players 2 to 4

Skill Multiplication facts 0 to 10

Object of the game To collect the most cards.

Directions

1. Shuffle the cards. Place the deck number-side down on the table.
2. Each player turns over 2 cards and calls out the product of the numbers.
3. The player with the largest product wins the round and takes all the cards.
4. In case of a tie for the largest product, each tied player turns over 2 more cards and calls out the product of the numbers. The player with the largest product then takes all the cards from both plays.
5. The game ends when there are not enough cards left for each player to have another turn.
6. The player with the most cards wins.

Example Ann turns over a 2 and a 6. She calls out 12.
Beth turns over a 6 and a 0. She calls out 0.
Joe turns over a 10 and a 4. He calls out 40.
Joe has the largest product. He takes all 6 cards.

Example Ann turns over a 3 and an 8.

3	8

She multiplies 3×8 and calls out 24.

Beth turns over a 4 and a 6.

4	6

She multiplies 4×6 and calls out 24.

Joe turns over a 9 and a 2.

9	2

He multiplies 9×2 and calls out 18.

Ann and Beth are tied with 24.
So they each turn over 2 more cards.

Ann turns over a 3 and a 7.

3	7

She multiplies 3×7 and calls out 21.

Beth turns over an 8 and a 4.

8	4

She multiplies 8×4 and calls out 32.
Beth wins and takes all 10 cards.

Name That Number

Materials ☐ number cards 0–20 (4 of each card 0–10, and 1 of each card 11–20)

Players 2 to 4 (the game is more interesting when played by 3 or 4 players)

Skill Naming numbers with expressions

Object of the game To collect the most cards.

Directions

1. Shuffle the deck and place 5 cards number-side up on the table. Leave the rest of the deck number-side down. Then turn over the top card of the deck and lay it down next to the deck. The number on this card is the number to be named. Call this number the **target number.**
2. Players take turns. When it is your turn:
   - Try to name the target number. You can name the target number by adding, subtracting, multiplying, or dividing the numbers on 2 or more of the 5 cards that are number-side up. A card may be used only once for each turn.
   - If you can name the target number, take the cards you used to name it. Also take the target-number card. Then replace all the cards you took by drawing from the top of the deck.
   - If you cannot name the target number, your turn is over. Turn over the top card of the deck and lay it down on the target-number pile. The number on this card becomes the new target number to be named.
3. Play continues until all of the cards in the deck have been turned over. The player who has taken the most cards wins.

Example Mae and Mike take turns.

4 10 8 12 2 [deck] 6

It is Mae's turn. The target number is 6. Mae names the number with $4 + 2$. She also could have said $8 - 2$ or $10 - 4$.

Mae takes the 4, 2, and 6 cards. Then she replaces them by drawing cards from the deck.

7 10 8 12 1 [deck] 16

It is Mike's turn. The new target number is 16. Mike sees two ways to name the target number.

- He can use 3 cards and name the target number like this:

$7 + 8 + 1 = 16$

- He can use 4 cards and name the target number like this:

$12 - 10 = 2$

$2 \times 8 = 16$

$16 \div 1 = 16$

Mike chooses the 4-card solution because he takes more cards that way. He takes the 12, 10, 8, and 1 cards. He also takes the target-number card 16. Then he replaces all 5 cards by drawing cards from the deck.

Number-Grid Difference

Materials
- ☐ 0–9 number cards (4 of each)
- ☐ 1 Completed Number Grid (*Math Masters,* p. 396)
- ☐ 1 *Number-Grid Difference* Record Sheet (*Math Masters,* p. 452)
- ☐ 2 counters
- ☐ 1 calculator

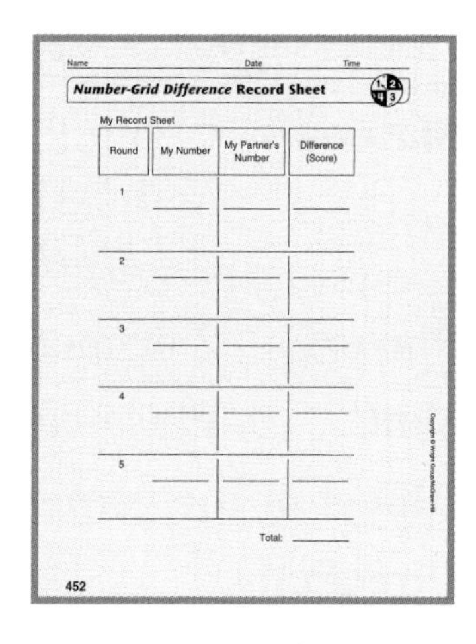

Name Date Time

***Number-Grid Difference* Record Sheet**

My Record Sheet

Round	My Number	My Partner's Number	Difference (Score)
1	______	______	______
2	______	______	______
3	______	______	______
4	______	______	______
5	______	______	______

Total: ______

452

Players 2

Skill Mental subtraction of 2-digit numbers

Object of the game To have the lower sum.

Directions

1. Shuffle the cards. Place the deck number-side down on the table.
2. Players take turns. When it is your turn:
   - ◆ Each player takes 2 cards from the deck and uses their cards to make a 2-digit number. Players then place counters on the grid to mark their numbers.
   - ◆ Find the difference between the 2 marked numbers.
   - ◆ This difference is your score for the turn. Record the 2 numbers and your score on the record sheet.
3. Continue playing until each player has taken 5 turns and recorded 5 scores.
4. Each player finds the sum of their 5 scores. Players may use a calculator to add.
5. The player with the lower sum wins the game.

Number Top-It (5-Digit Numbers)

Materials ☐ number cards 0–9 (4 of each)
☐ 1 7-Digit Place-Value Mat
(*Math Masters,* pp. 423 and 424)

Players 2 or more

Skill Place value for whole numbers

Object of the game To make the largest 5-digit numbers.

Directions

1. Shuffle the cards. Place the deck number-side down on the table.
2. Each player uses 1 row of boxes on the Place-Value Mat. Do not use the Millions box or the Hundred-Thousands box.
3. In each round, players take turns turning over the top card from the deck and placing it on any one of their empty boxes. Each player takes 5 turns and places 5 cards on his or her row of the Place-Value Mat.
4. At the end of each round, players read their numbers aloud and compare them. The player with the largest number for the round scores 1 point; the player with the next-largest number scores 2 points; and so on. All cards are then removed from the Place-Value Mat and placed in a discard pile before the next round begins.
5. Players play 5 rounds per game. When all of the cards in the deck have been used, one player shuffles the discarded cards to make a new deck to finish the game. The player with the smallest total number of points at the end of 5 rounds wins the game.

Example The Place-Value Mat below shows the results for one complete round of play with 4 players.

7-Digit Place-Value Mat

	Millions	Hundred Thousands	Ten Thousands	Thousands	Hundreds	Tens	Ones
John			4	8	6	2	1
Doug			9	3	5	2	0
Sara			4	7	2	0	4
Anju			7	6	6	3	4

Here are the numbers listed from largest to smallest:

Doug	93,520	largest
Anju	76,634	
John	48,621	
Sara	47,204	smallest

Doug scores 1 point for this round. Anju scores 2 points. John scores 3 points. And Sara scores 4 points.

Number Top-It (7-Digit Numbers)

Materials ☐ number cards 0–9 (4 of each)
☐ 1 7-Digit Place-Value Mat (*Math Masters,* pp. 423 and 424)

Players 2 or more

Skill Place value for whole numbers

Object of the game To make the largest 7-digit numbers.

Directions

This game is played in the same way as *Number Top-It* (5-Digit Numbers). The only difference is that each player uses all 7 boxes in one row of the Place-Value Mat.

In each round, Players take turns turning over the top card from the deck and placing it on any one of their empty boxes. Each player takes 7 turns and places 7 cards on his or her row of the game mat.

Example Andy and Barb played 7-digit *Number Top-It.* Here is the result for one complete round of play:

7-Digit Place-Value Mat

	Millions	Hundred Thousands	Ten Thousands	Thousands	Hundreds	Tens	Ones
Andy	7	6	4	5	2	0	1
Barb	4	9	7	3	5	2	4

Andy's number is larger than Barb's number. Andy scores 1 point for the round. Barb scores 2 points.

Number Top-It (Decimals)

Materials ☐ number cards 0–9 (4 of each)
☐ 1 *Number Top-It* Mat (Decimals) (*Math Masters,* pp. 453 or 454)

Players 2 or more

Skill Place value for decimals

Object of the game To make the largest 2-digit decimal numbers.

Directions

This game is played in the same way as *Number Top-It* (5-Digit Numbers). The only difference is that players use a *Number Top-It* Mat for decimals.

In each round, players take turns turning over the top card from the deck and placing it on any one of their empty boxes. Each player takes 2 turns and places 2 cards on his or her row of the game mat.

Example Andy and Barb played *Number Top-It* using the *Number Top-It* Mat (2-Place Decimals). Here is the result:

Barb's number is larger than Andy's number. Barb scores 1 point for the round. Andy scores 2 points.

***Number Top-It* Mat (2-Place Decimals)**

	Ones	.	Tenths	Hundredths
Andy	0	.	3	5
Barb	0	.	6	4

You may also use a *Number Top-It* Mat that has empty boxes in the tenths, hundredths, and thousandths places. Each player takes 3 turns and places 3 cards.

Pick-a-Coin

Materials
- ☐ 1 six-sided die
- ☐ 1 calculator for each player
- ☐ 1 *Pick-a-Coin* Record Table (*Math Masters,* p. 455)

Players 2 or 3

Skill Place value for decimals

Object of the game To make the largest dollar-and-cents amounts.

Directions

Each player uses a different Record Table. Players take turns. When it is your turn:

1. Roll the die 5 times. After each roll, write the number that comes up in any one of the empty cells on a line of your Record Table.
2. Use a calculator to find the total amount for that turn.
3. Record the total for that turn on your Record Table.
4. After 4 turns, use your calculator to add the 4 totals. The player with the largest sum wins.

Example On his first turn, Brian rolled 4, 2, 4, 6, and 1. He filled in his Record Table like this.

***Pick-a-Coin* Record Table**

Brian	Ⓟ	Ⓝ	Ⓓ	Ⓠ	$1	Total
1st turn	2	1	4	4	6	$ 7.47
2nd turn						$ ___.___
3rd turn						$ ___.___
4th turn						$ ___.___
					Total	$ ___.___

Roll to 100

Materials ☐ 1 *Roll to 100* Record Sheet (*Math Masters,* p. 456)
☐ 2 six-sided dice

Players 2 to 4

Skill Mental addition skills; developing a winning game strategy

Object of the game To score at least 100.

Directions

Players take turns.

1. When it is your first turn, roll the dice any number of times.
   - ◆ Mentally add all of the numbers rolled for all your dice rolls. Enter this as your score for Turn 1.
   - ◆ If you roll a 1 at any time, your turn is over. Enter 0 as your score for Turn 1.
2. When you take another turn, roll the dice any number of times.
   - ◆ Start with your score from the last turn and keep mentally adding on all of the numbers you roll. Do this until you stop rolling the dice. Enter your final sum as your score for this turn.
   - ◆ If you roll a 1 at any time, your turn is over. The score you enter for this turn is the same as your score on the previous turn.
3. The first player to score 100 or more wins the game.

Variations

Double the Doubles: Whenever a player rolls doubles, the numbers rolled are each added twice.

Double Ones: A player who rolls double 1s enters a score of 0 for that turn. On that player's next turn, he or she starts with a score of 0.

Back to Zero: A player who scores 100 or more continues to take turns, but subtracts the numbers rolled each time instead of adding them. The first player to score 100 or more and then get back to 0 or less wins.

Name ______ Date ______ Time ______

***Roll to 100* Record Sheet**

Write your score at the end of each turn. The first player to reach or pass 100 wins.

Turn	Player 1	Player 2	Player 3	Player 4
1				
2				
3				
4				
5				
6				
7				
8				
9				
10				

Continue recording scores on the back of this page.

456

Spinning to Win

Materials ☐ 1 paper clip—preferably large size (2 in.)
☐ 50 counters
☐ 1 spinner (*Math Masters,* p. 464)

Players 2 to 4

Skill Using chance data to develop a winning game strategy

Object of the game To collect the most counters in 12 spins.

Directions

Win 1	Win 2	Win 5	Win 10

1. Put the counters in a pile on the table between the players.
2. For each game, draw a tally chart like the one at the right.
3. Each player claims one section of the spinner—1, 2, 5, or 10. Each player must choose a different section.
4. Players take turns spinning the paper clip. One game consists of a total of 12 spins.
5. When the paper clip lands on a section of the spinner that has been claimed, the player that claimed the section takes the number of counters printed there. Make a tally mark in the chart to show the winning number for that spin. Do this to keep track of the number of spins.
6. The winner is the player with the most counters after 12 spins.

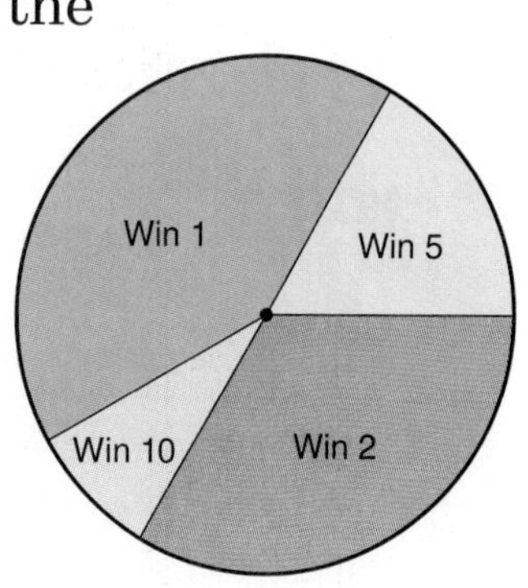

Subtraction Top-It

Materials ☐ number cards 0–20 (4 of each card 0–10, and 1 of each card 11–20)

Players 2 to 4

Skill Subtraction facts

Object of the game To collect the most cards.

Directions

1. Shuffle the cards. Place the deck number-side down on the table.
2. Each player turns over 2 cards and subtracts the smaller number from the larger number.
3. The player with the largest difference wins the round and takes all the cards.
4. In case of a tie for the largest difference, each tied player turns over 2 more cards and calls out the difference of the numbers. The player with the largest difference then takes all the cards from both plays.
5. The game ends when not enough cards are left for each player to have another turn.
6. The player with the most cards wins.

Example Ann turns over a 2 and a 14.
She subtracts 2 from 14 and calls out 12.

2 14

Joe turns over a 10 and a 4.
He subtracts 4 from 10 and calls out 6.

10 4

Ann has the larger difference. She takes all 4 cards.

Example Ann turns over a 12 and a 6.

12 6

She subtracts 12 − 6 and calls out 6.

Joe turns over a 9 and a 3.

9 3

He subtracts 9 − 3 and calls out 6.

There is a tie. So both players turn over 2 more cards.

Ann turns over a 10 and an 8.

10 8

She subtracts 10 − 8 and calls out 2.

Joe turns over a 7 and a 3.

7 3

He subtracts 7 − 3 and calls out 4.
Joe takes all 8 cards.

Target 50

Materials
- ☐ number cards 0–9 (4 of each)
- ☐ base-10 blocks (30 longs and 30 cubes)
- ☐ 1 Place-Value Mat for each player (*Math Masters,* p. 411)
- ☐ 1 *Target 50* Record Sheet (*Math Masters,* p. 465)

Players 2

Skill Place value for whole numbers

Object of the game To have 5 longs on the Place-Value Mat.

Directions

1. Shuffle the number cards. Place the deck number-side down on the table.
2. Players take turns. When it is your turn:
   - Turn over 2 cards. You may use either card to make a 1-digit number. Or, you may use both cards to make a 2-digit number.
   - Use base-10 blocks to model your number. Put these blocks just beneath your Place-Value Mat, but not on the mat.
   - You now have 2 choices:

     Addition: You can add all of the base-10 blocks beneath the mat to the blocks already on your Place-Value Mat.

     Subtraction: You can subtract blocks equal in value to the base-10 blocks beneath the mat from the blocks already on your Place-Value Mat. If you decide to subtract, you may have to make exchanges on the Place-Value Mat first.

3. Players can make exchanges on their Place-Value Mats at any time.

4. Play continues until the blocks on one player's mat have a value of 50 and show 5 longs. That player is the winner.

Example Alex was able to reach the target value of 50 in three turns:

Turns	Cards	Number Made	Addition or Subtraction on Place-Value Mat	Value on Mat
1	6, 5	56	**Add** 5 longs and 6 cubes.	56
2	8, 9	8	Exchange 1 long for 10 cubes **Subtract** 8 cubes.	48
3	5, 2	2	**Add** 2 cubes. Exchange 10 cubes for 1 long.	50

Name Date Time

***Target: 50* Record Sheet**

For each of your turns, record the number you make and the value you show with base-10 blocks on the Place-Value Mat.

Turns	Number You Made	Value on Place-Value Mat
1		
2		
3		
4		
5		
6		
7		
8		
9		
10		

465

Three Addends

Materials ☐ number cards 0–20 (4 of each card 0–10, and 1 of each card 11–20)

☐ 1 *Three Addends* Record Sheet for each player (*Math Masters,* p. 466, optional)

Players 2

Skill Addition of three 1- and 2-digit numbers

Object of the game To find easy combinations when adding three numbers.

Directions

1. Shuffle the cards and place the deck number-side down on the table.
2. One player draws 3 cards from the top of the deck and turns them over.
3. Each player writes an addition number model using the 3 numbers.
4. You can write your addition number model on your Record Sheet or on a separate sheet of paper.
5. You may list the numbers in any order you wish. But try to list the numbers so that it is easy for you to add them.
6. Then add the numbers and compare your answer to the other player's answer.

2 7 8

$2 + 8 + 7 = 17$

Variations

- Give the sum of the 3 card numbers without writing down a number model.
- Draw 4 cards from the deck. Turn them over and find the sum of the 4 numbers.

A.M. An abbreviation that means "before noon." It refers to the period between midnight (12 A.M.) and noon (12 P.M.).

Angle A figure that is formed by two rays or two line segments that have the same endpoint.

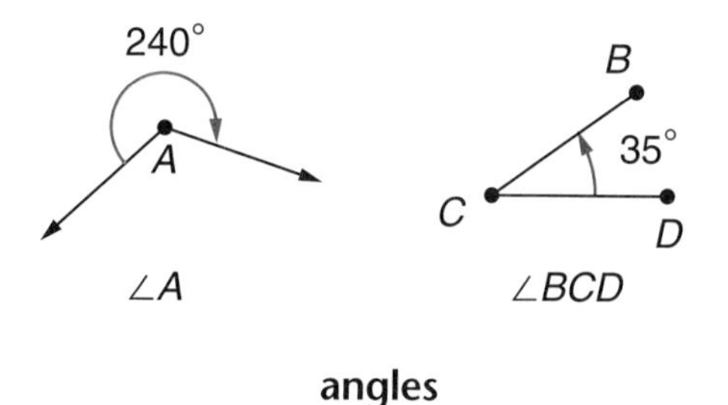

angles

Apex In a pyramid or a cone, the vertex opposite the *base*. In a pyramid, all the faces except the base meet at the apex.

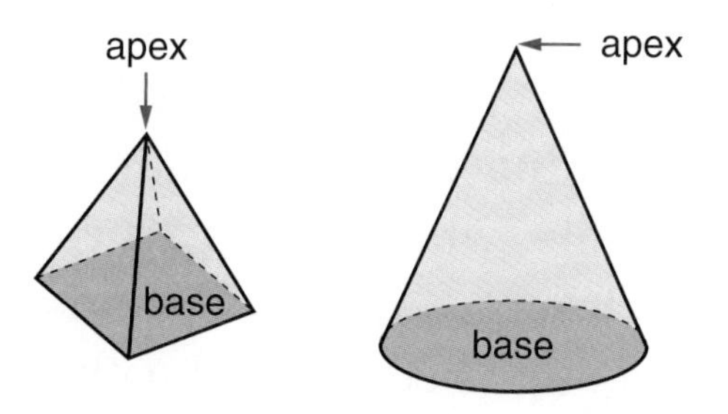

Area The amount of surface inside a shape. Area is measured in square units, such as square inches or square centimeters.

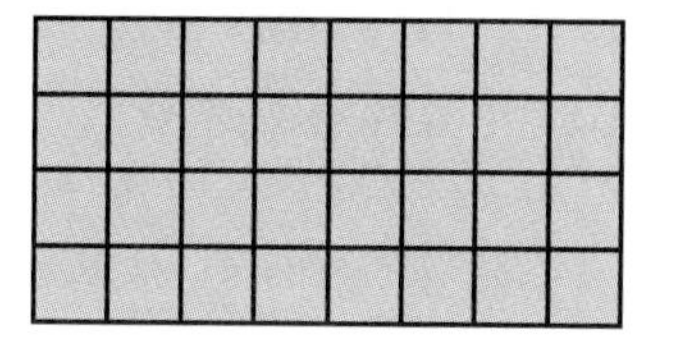

32 square units

about 21 square units

1 square centimeter

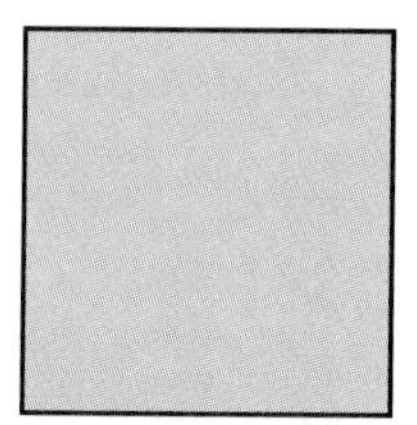

1 square inch

Array An arrangement of objects into rows and columns that form a rectangle. All rows and columns must be filled. Each row has the same number of objects. And each column has the same number of objects.

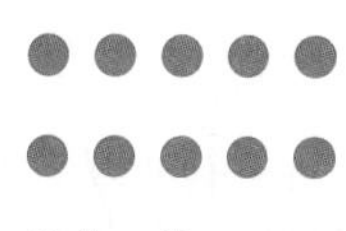

2-by-5 array
(2 rows, 5 columns)

Average See *mean*.

Ballpark estimate A rough estimate to help you solve a problem or check an answer.

Bar graph A graph that uses horizontal or vertical bars to represent data.

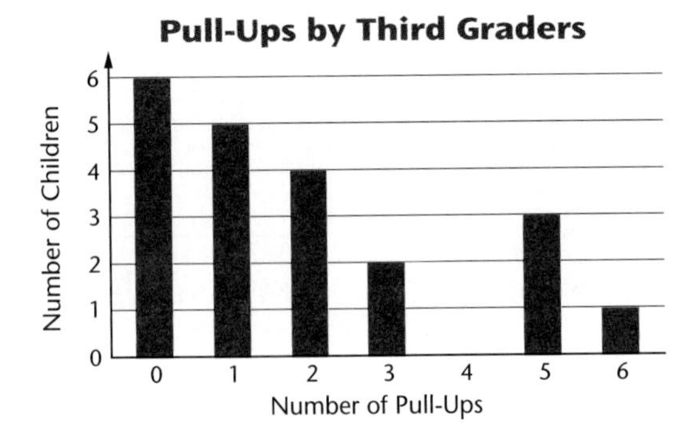

Base A name used for a side of a polygon or a face of a 3-dimensional figure.

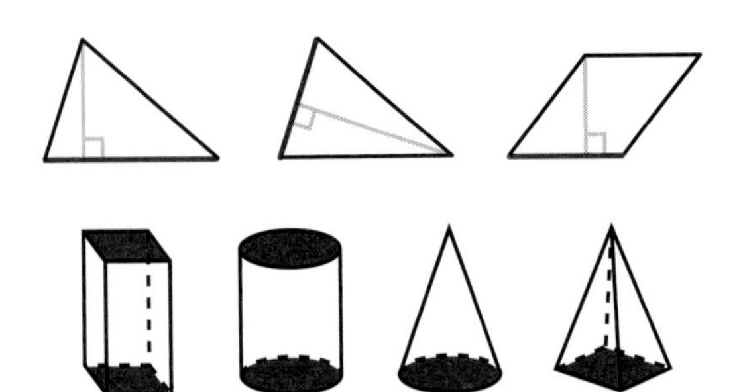

Bases are shown in red.

Capacity (1) The amount a container can hold. The *volume* of a container. Capacity is usually measured in units such as gallons, pints, cups, and liters. (2) The heaviest weight a scale can measure.

Celsius The temperature scale used in the metric system.

Center The point inside a circle or sphere that is the same distance from all of the points on the circle or sphere.

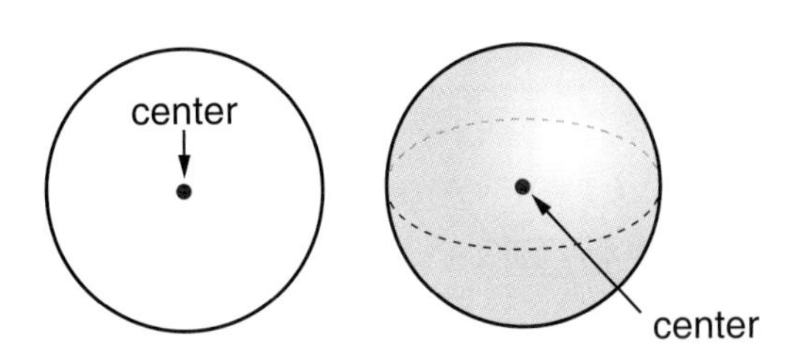

Chance The possibility that something will occur. For example, when you flip a coin, the chance that it will land heads up is equal to the chance that it will land tails up.

Change number story A number story in which an amount is increased (a change-to-more story) or decreased (a change-to-less story). A change diagram can be used to keep track of the numbers and missing information in such problems.

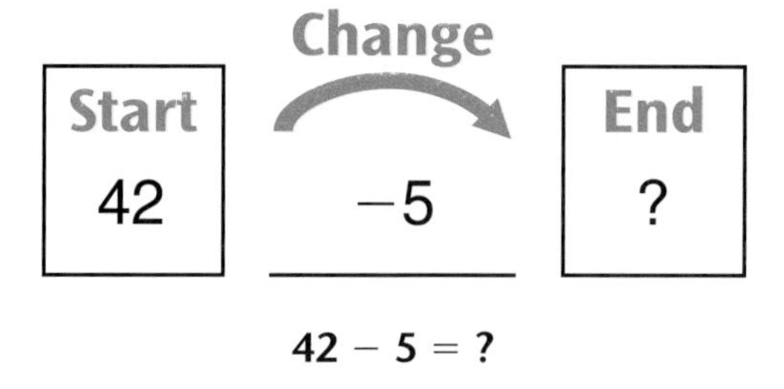

Circle A curved line that forms a closed path on a flat surface so that all points on the path are the same distance from a point called the *center*.

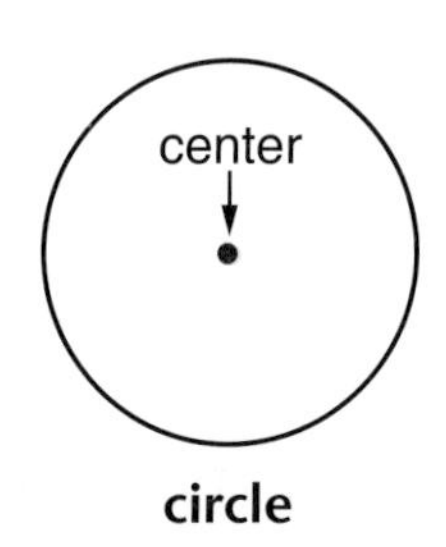

Circumference The distance around a circle; the *perimeter* of a circle.

Comparison number story A number story in which two quantities are compared. A comparison diagram can be used to keep track of the numbers and missing information in such problems.

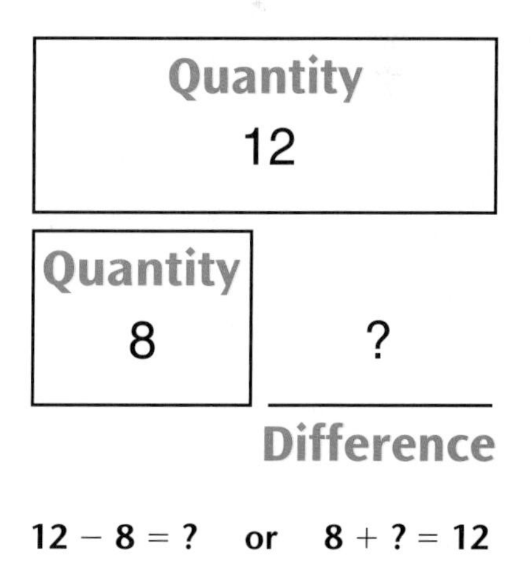

Composite number A counting number that has more than two different *factors*. For example, 4 is a composite number because it has three factors: 1, 2, and 4.

Cone A solid that has a circular *base* and a curved surface that ends at a point called the *apex*.

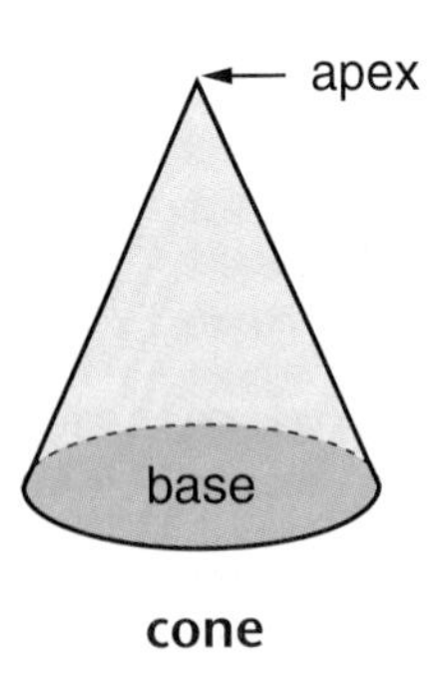

cone

Congruent figures Figures that have the same shape and the same size. Two figures on a flat surface are congruent if they match exactly when one is placed on top of the other.

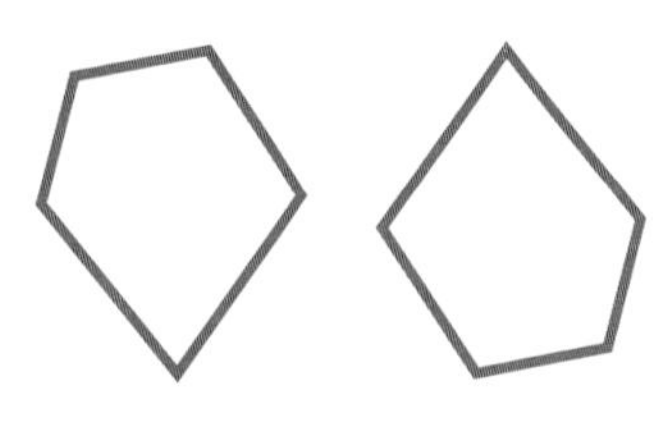
congruent pentagons

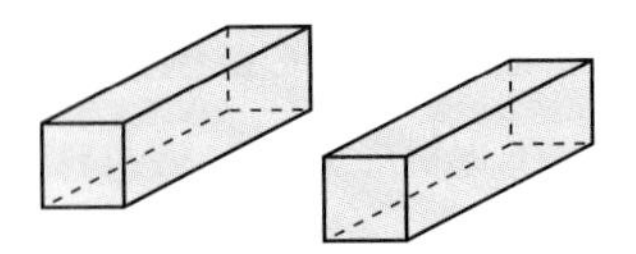
congruent prisms

Coordinate grid A grid formed by drawing two number lines that form right angles. The number lines intersect at their zero points. You can use *ordered pairs* of numbers to locate points on the grid. (The numbers in each pair are called *coordinates*.) Maps are often based on coordinate grids.

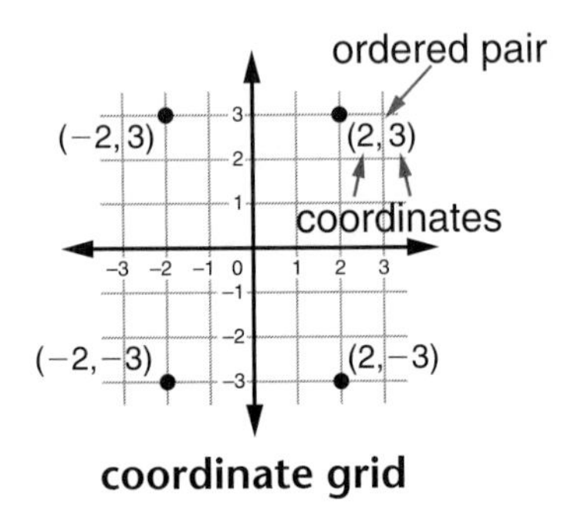

coordinate grid

Coordinates See *ordered pair*.

Counting numbers The numbers used in counting: 1, 2, 3, 4, and so on. Zero is sometimes thought of as a counting number.

Cylinder A solid that has two circular *bases* that are parallel and the same size. The bases are connected by a curved surface. A soup can is shaped like a cylinder.

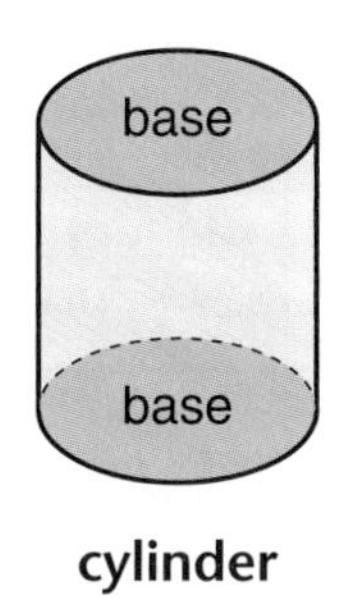

cylinder

Data Information that is collected by counting, measuring, asking questions, or observing.

Decimal A number, such as 23.4, that contains a *decimal point*. Money amounts, such as \$6.58, are decimal numbers. The decimal point in money separates the dollars from the cents.

Decimal point A dot used to separate the ones place from the tenths place in decimal numbers.

Degree (°) (1) A unit of measure for angles. (2) A unit of measure for temperature. In both cases, a small raised circle (°) is used to show degrees.

Denominator The number below the line in a fraction. For example, in $\frac{3}{4}$, 4 is the denominator.

Diameter (1) A line segment that goes through the center of a circle and has endpoints on the circle. (2) The length of this line segment. The diameter of a sphere is defined in the same way. The diameter of a circle or sphere is twice the length of its *radius*.

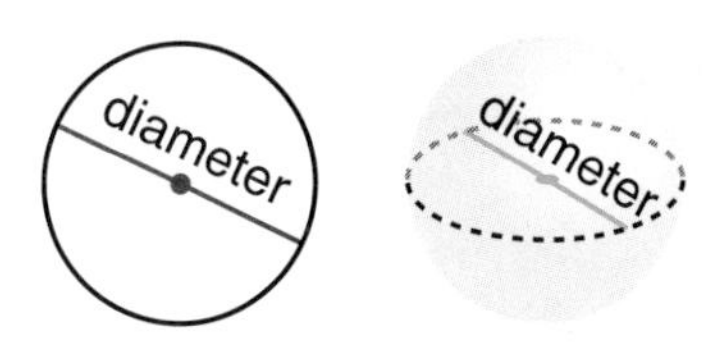

Digits The symbols 0, 1, 2, 3, 4, 5, 6, 7, 8, and 9 that are used to write any number in our number system.

Edge A line segment or curve where the surfaces of a solid meet.

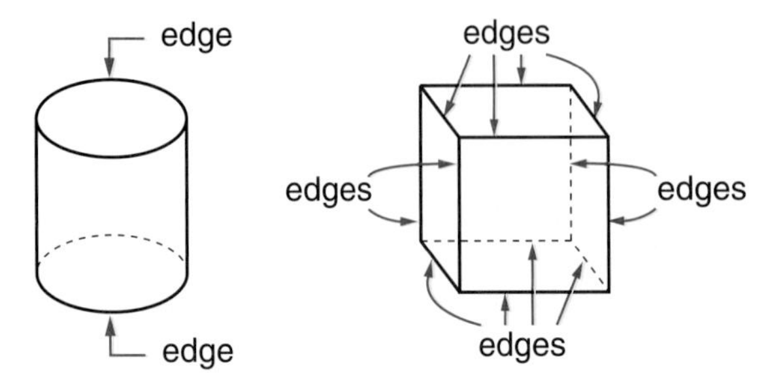

Endpoint A point at the end of a *line segment* or *ray*. A line segment is named using the letter labels of its endpoints. A ray is named using the letter labels of its endpoint and another point on the ray.

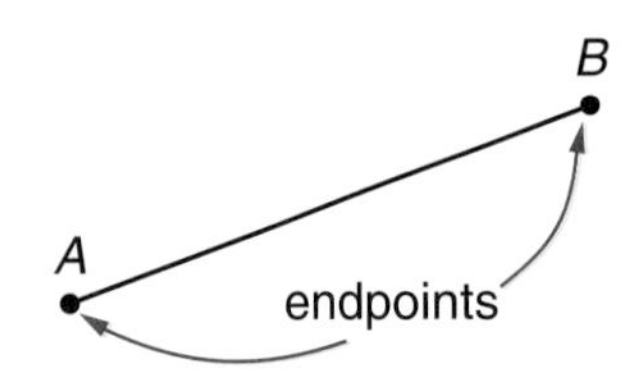

line segment *AB* or *BA*

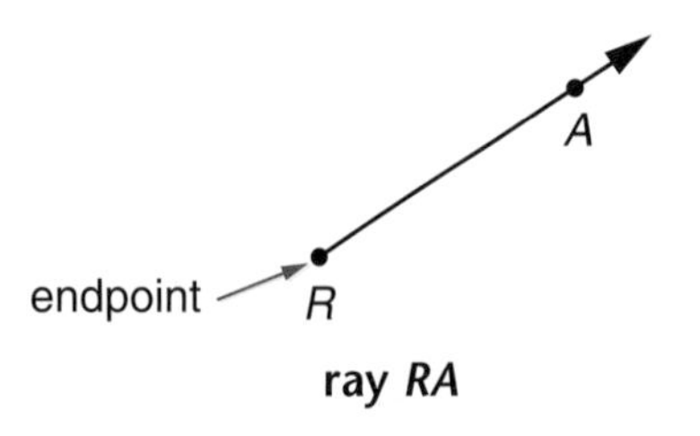

ray *RA*

Equal-grouping story A number story in which a total collection is separated into *equal groups.* A diagram can be used to keep track of the numbers and missing information in such problems.

teams	players per team	players in all
?	9	54

Each group has 9 players. How many groups if there are 54 players?

Equal groups Collections or groups of things that all contain the same number of things. For example, boxes that each contain 100 clips are equal groups. And rows of chairs with 6 chairs per row are equal groups.

Equal-sharing story A number story in which a group of things is divided into equal parts, called *shares*. A diagram can be used to keep track of the numbers and missing information in such problems.

children	apples per child	apples in all
4	?	24

If 24 apples are divided among 4 children into equal shares, how many apples are in each share?

Equally likely Two or more events that each have the same chance of happening. For example, when you roll a 6-sided die, each of the sides is equally likely to land faceup.

Equilateral triangle A triangle with all three sides equal in length. In an equilateral triangle, all three angles have the same measure.

equilateral triangle

Equivalent names Different ways to name the same number. For example, $2 + 6$, $12 - 4$, 2×4, $16 \div 2$, $5 + 1 + 2$, VIII, eight, and 𝍸/// are equivalent names for 8.

Estimate An answer that should be close to an exact answer. *To estimate* means to give an answer that should be close to an exact answer.

Even number A counting number that can be divided by 2 with no remainder. The even numbers are 2, 4, 6, 8, and so on.

Event Something that happens. Tossing heads with a coin is an event. Rolling a number smaller than 5 with a die is an event. The *probability* of an event is the chance that the event will happen.

Face A flat surface on the outside of a solid.

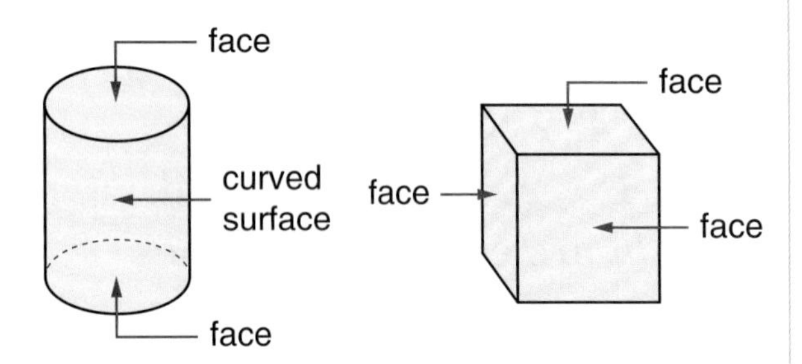

Fact family (1) A set of related addition and subtraction facts. For example, $5 + 6 = 11$, $6 + 5 = 11$, $11 - 5 = 6$, and $11 - 6 = 5$ are a fact family. (2) A set of related multiplication and division facts. For example, $5 \times 7 = 35$, $7 \times 5 = 35$, $35 \div 5 = 7$, and $35 \div 7 = 5$ are a fact family.

Fact Triangles Cards with a triangle shape that show *fact families.* Fact Triangles are used like flash cards to help you memorize basic addition, subtraction, multiplication, and division facts.

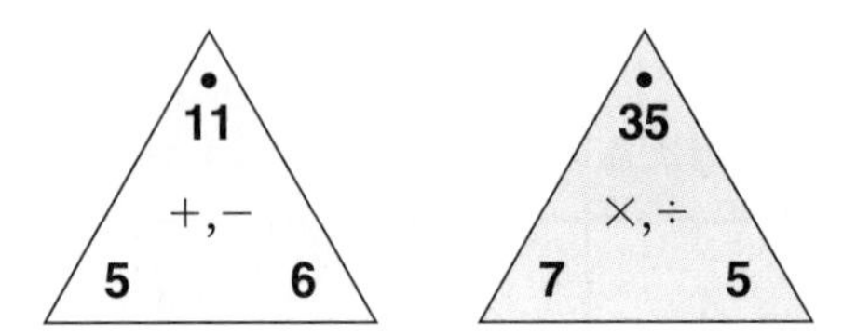

Factor (1) Any of the numbers that are multiplied to find a product. For example, in the problem $4 \times 7 = 28$, 28 is the product, and 4 and 7 are the factors. (2) A number that divides another number evenly. For example, 8 is a factor of 24 because $24 \div 8 = 3$, with no remainder.

Facts Table A chart with rows and columns that shows all of the basic addition and subtraction facts, or all of the basic multiplication and division facts.

Fahrenheit The temperature scale used in the U.S. customary system.

Fraction A number in the form $\frac{a}{b}$ or a/b. The number a is called the *numerator* and can be any counting number or 0. The number b is called the *denominator* and can be any counting number except 0. One use for fractions is to name part of a whole or part of a collection.

Frames and Arrows A diagram used in *Everyday Mathematics* to show a number pattern or sequence.

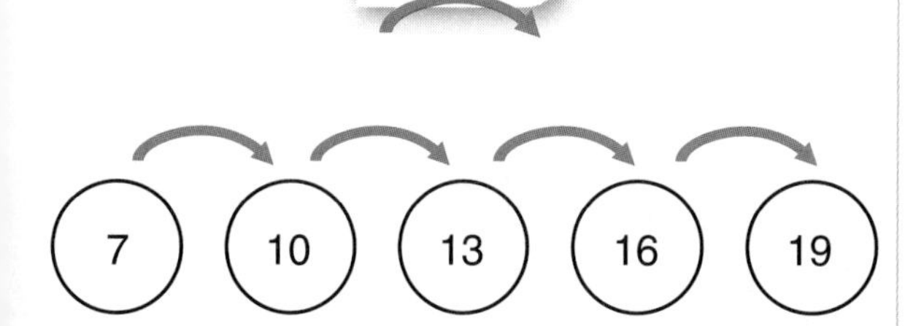

Function machine An imaginary machine used in *Everyday Mathematics* to change numbers according to a given rule.

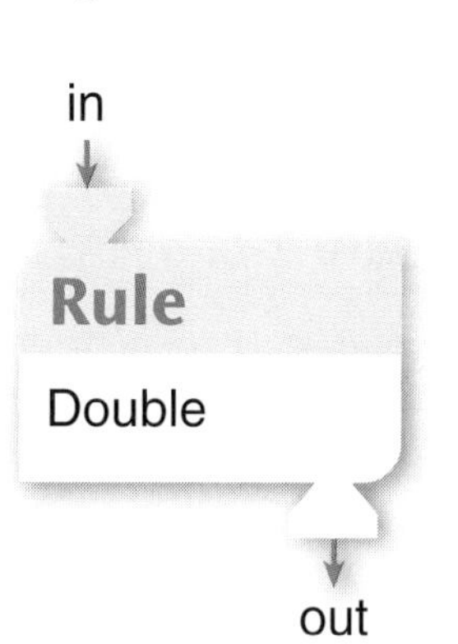

in	out
1	2
3	6
5	10
10	20
100	200

Geometry The study of shapes.

Intersect To meet or cross.

intersecting segments

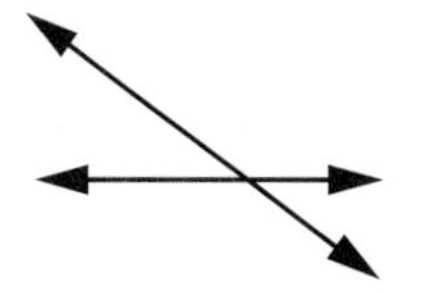

intersecting lines

Kite A 4-sided polygon with two pairs of equal sides. The equal sides are next to each other. The four sides cannot all have the same length. (So a rhombus is not a kite.)

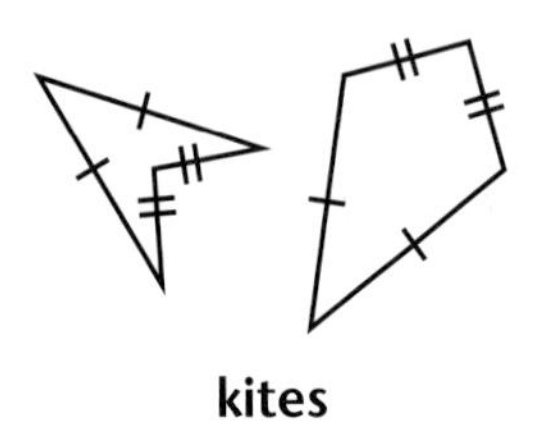

kites

Lattice method One method for solving multiplication problems.

Line A straight path that goes on forever in both directions.

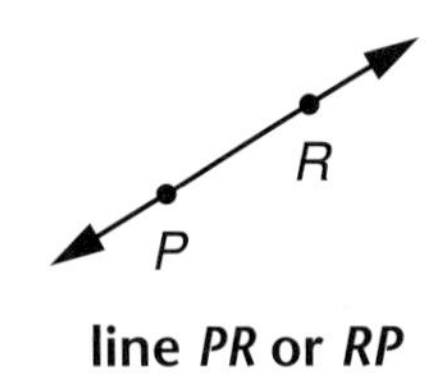

line *PR* or *RP*

Line graph A graph that uses line segments to connect data points. Line graphs are often used to show how something has changed over a period of time.

Outdoor Temperature at 2:00 P.M.

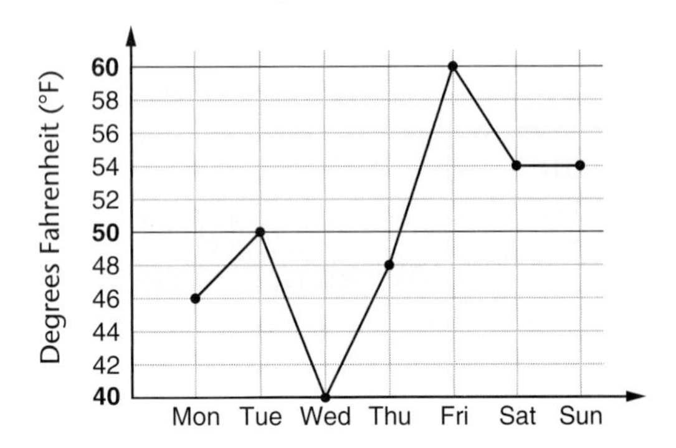

Line plot A sketch of data that uses Xs, checks, or other marks above a number line to show how many times each value appears in the set of data.

Test Scores

Number of Children						
					X	
					X	
				X	X	
			X	X	X	
		X	X	X	X	
		X	X	X	X	X
	0	1	2	3	4	5

Number Correct

Line segment A straight path joining two points. The two points are called *endpoints* of the segment.

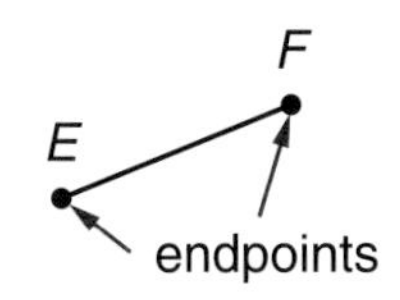

line segment *EF* or *FE*

Line symmetry A figure has line symmetry if a line can divide it into two parts that look like mirror images of each other. The two parts look alike but face in opposite directions. The dividing line is called the *line of symmetry.*

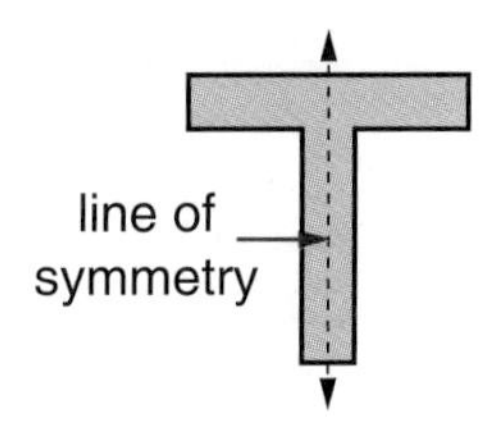

M

Maximum The largest amount. The largest number in a set of data.

Mean The *average* number in a set of data. The mean is found by adding all of the data values and then dividing by the number of numbers in the set of data.

Median The middle number in a set of data when the numbers are put in order from smallest to largest, or from largest to smallest. The median is also known as the *middle number* or *middle value.*

Metric system A measuring system that is used by scientists everywhere, and in most countries in the world except the United States. The metric system is a decimal system. It is based on multiples of 10. See the Tables of Measures on pages 246 and 247.

Minimum The smallest amount. The smallest number in a set of data.

Mode The number or value that occurs most often in a set of data.

Name-collection box In *Everyday Mathematics,* a place to write *equivalent names* for the same number.

50

$100 \div 2$ 5×10

$10 + 10 + 10 + 10 + 10$

1 more than 49 $25 + 25$

fifty *cincuenta*

name-collection box

Negative number A number that is less than zero. A number to the left of zero on a horizontal number line. A number below zero on a vertical number line. The symbol $-$ may be used to write a negative number. For example, "negative 5" is usually written as -5.

Number grid A table with rows and columns that lists numbers in order. A monthly calendar is a number grid.

Number line A line with numbers marked in order on it.

Number model A group of numbers and symbols that shows how a number story can be solved. For example, $10 - 6 = 4$ and $10 - 6$ are each number models for the following story:

I had 10 cookies. I gave 6 away. How many did I have left?

Numerator The number above the line in a fraction. For example, in $\frac{3}{4}$, 3 is the numerator.

Odd number A counting number that cannot be exactly divided by 2. When an odd number is divided by 2, there is a remainder of 1. The odd numbers are 1, 3, 5, and so on.

Ordered pair A pair of numbers, such as (5,3) or (1,4), used to find a location on a coordinate grid. The numbers in an ordered pair are called *coordinates*. See *coordinate grid* for a diagram.

Parallel Always the same distance apart, and never meeting or crossing each other, no matter how far extended. Line segments are parallel if they are parts of lines that are parallel. The bases of a prism are parallel. The bases of a cylinder are parallel.

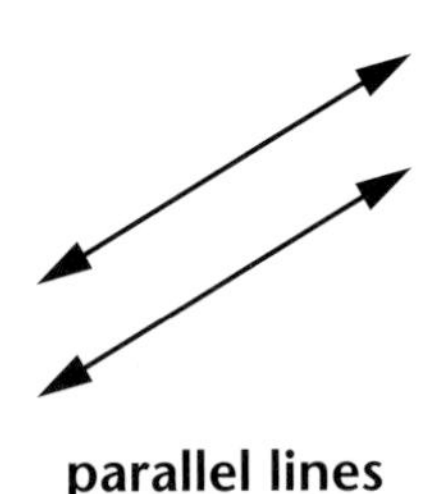

parallel lines

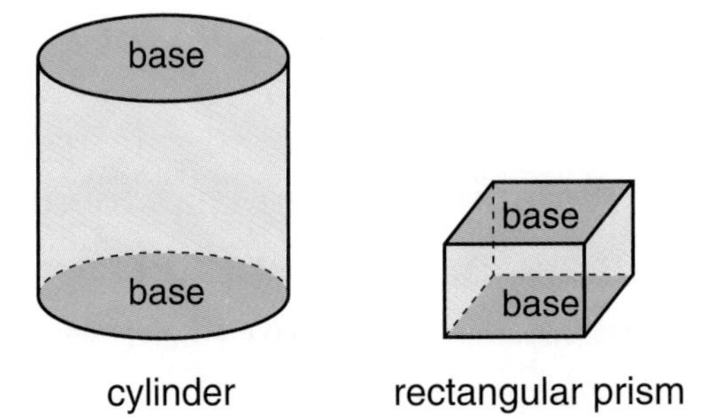

parallel bases

Parallelogram A 4-sided *polygon* whose opposite sides are parallel. The opposite sides of a parallelogram are also the same length. And the opposite angles in a parallelogram have the same measure.

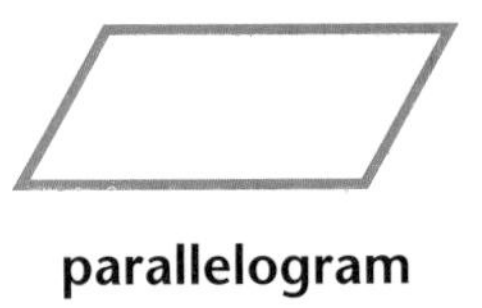

parallelogram

Partial-products method One method for solving multiplication problems.

Partial-sums method One method for solving addition problems.

Parts-and-total number story A number story in which two parts are combined to find a total. A parts-and-total diagram can be used to keep track of the numbers and missing information in such problems.

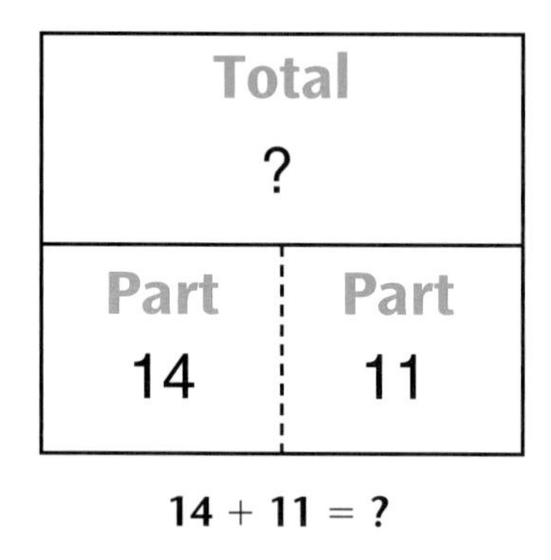

14 + 11 = ?

Pattern Shapes or numbers that repeat in a regular way so that what comes next can be predicted.

Per "For each" or "in each." For example, "three tickets per student" means "three tickets for each student."

Perimeter The distance around a polygon or other shape. The perimeter of a circle is called its *circumference.*

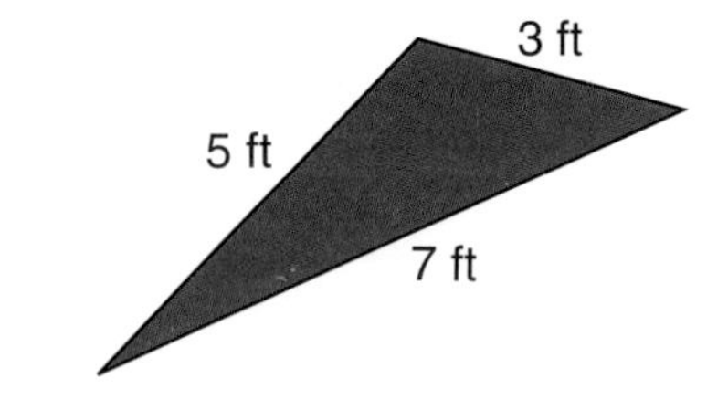

perimeter = 5 ft + 3 ft + 7 ft = 15 ft

Pictograph A graph that uses pictures or symbols to show numbers. The *key* for a pictograph tells what each picture or symbol is worth.

Number of Cars Washed

Friday
Saturday
Sunday

KEY: = 6 cars

Place value A system for writing numbers in which the value of a digit depends on its place in the number.

P.M. An abbreviation that means "after noon." It refers to the period between noon (12 P.M.) and midnight (12 A.M.).

Point An exact location in space.

Polygon A closed figure on a flat surface that is made up of line segments joined end to end. The line segments make one closed path and may not cross.

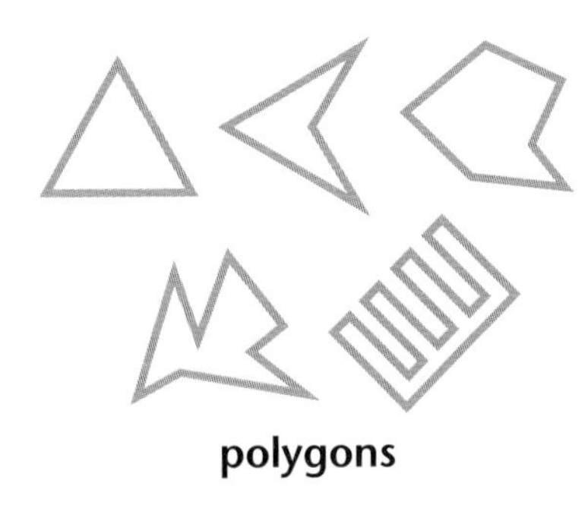

polygons

Polyhedron A solid whose surfaces (called *faces*) are all flat and formed by *polygons*. A polyhedron does not have any curved surfaces.

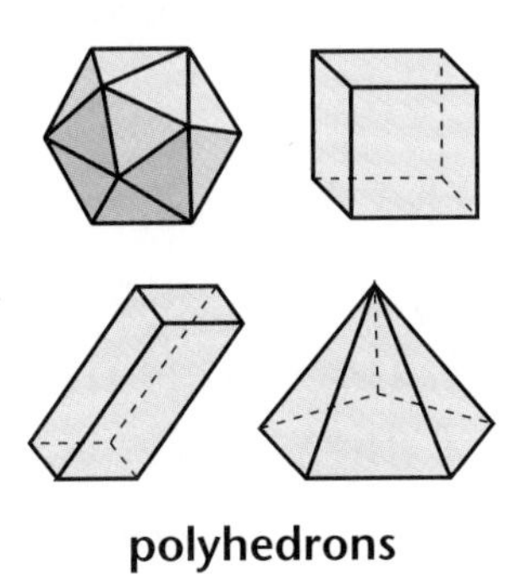

polyhedrons

Positive number A number that is greater than zero. A number to the right of zero on a horizontal number line. A number above zero on a vertical number line. A positive number may be written using the + symbol, but is usually written without it. For example, $+10 = 10$.

Prime number A counting number that has exactly two different *factors* that are counting numbers: itself and 1. For example, 5 is a prime number because its only factors are 5 and 1. The number 1 is not a prime number because that number has only a single factor, the number 1 itself.

Prism A *polyhedron* that has two parallel *bases* that are formed by polygons with the same size and shape. The other faces connect the bases and are all shaped like *parallelograms*. These other faces are often rectangles. Prisms take their names from the shape of their bases.

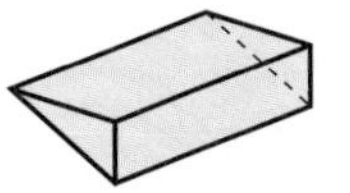

triangular prism

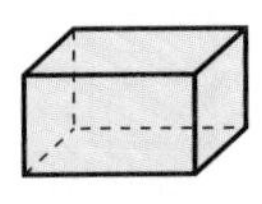

rectangular prism

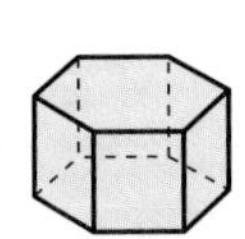

hexagonal prism

Probability A number from 0 through 1 that tells the *chance* that an *event* will happen. The closer a probability is to 1, the more likely the event is to happen.

Pyramid A *polyhedron* in which one face, the *base,* may have any polygon shape. All of the other faces have triangle shapes and come together at a vertex called the *apex.* A pyramid takes its name from the shape of its base.

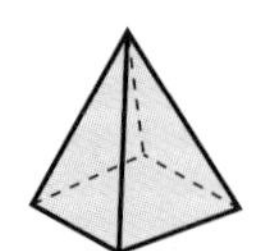
rectangular pyramid

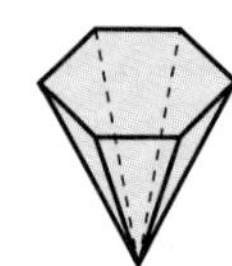
hexagonal pyramid

Q

Quadrangle A *polygon* that has four angles. Same as *quadrilateral.*

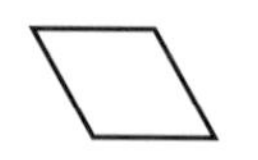
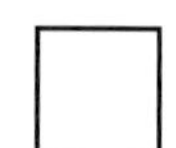

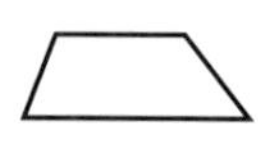

Quadrilateral A *polygon* that has four sides. Same as *quadrangle.*

R

Radius (plural: **radii**) (1) A line segment from the center of a circle to any point on the circle. (2) The length of this line segment. The radius of a sphere is defined in the same way. The radius of a circle or sphere is one-half the length of its *diameter.*

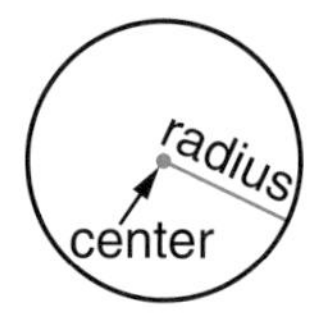

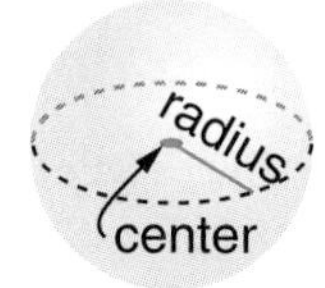

Range The difference between the largest (*maximum*) and the smallest (*minimum*) numbers in a set of data.

Ray A straight path that has one endpoint and goes on forever in one direction.

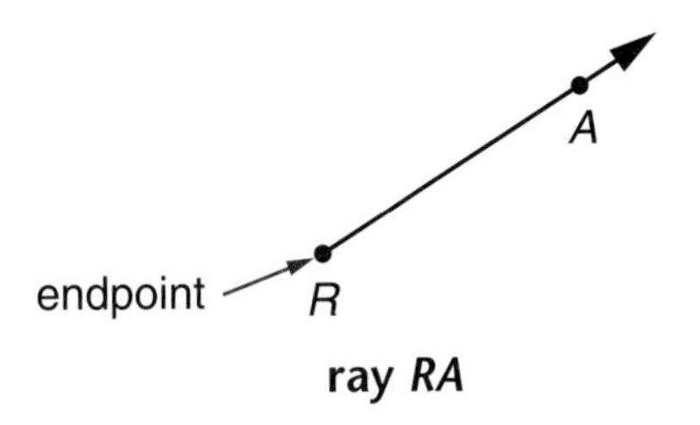

ray *RA*

Rectangle A *parallelogram* whose corners are all right angles.

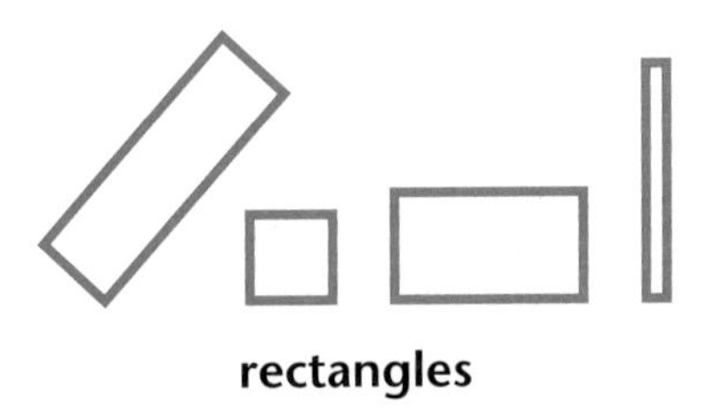
rectangles

Regular polygon A *polygon* whose sides all have the same length and whose angles (inside the polygon) all have the same size.

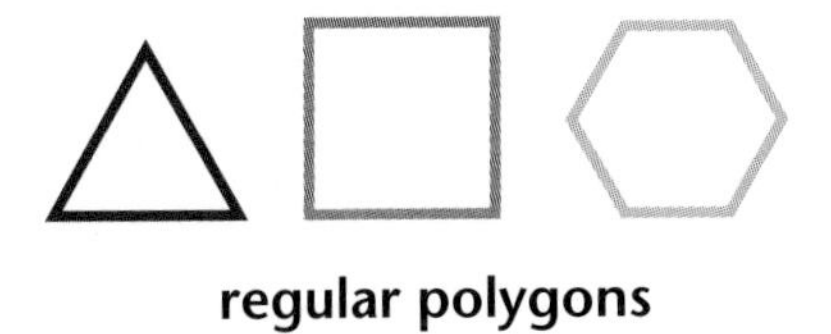
regular polygons

Remainder The amount left over when things are divided or shared equally. Sometimes there is no remainder.

Rhombus A *parallelogram* with all four sides the same length. Every square is a rhombus, but not all rhombuses are squares.

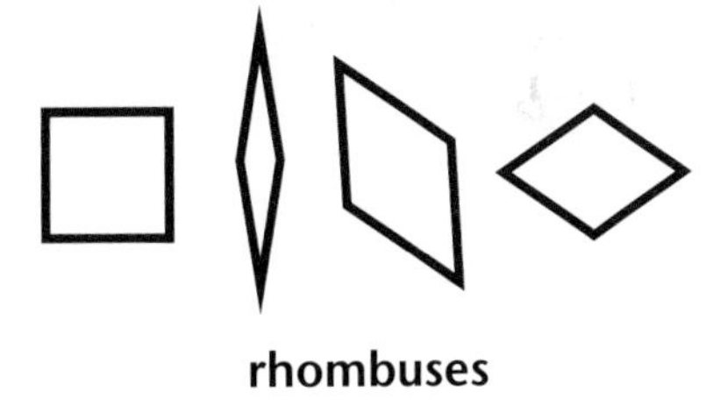
rhombuses

Right angle A 90° angle. The sides of a right angle form a square corner.

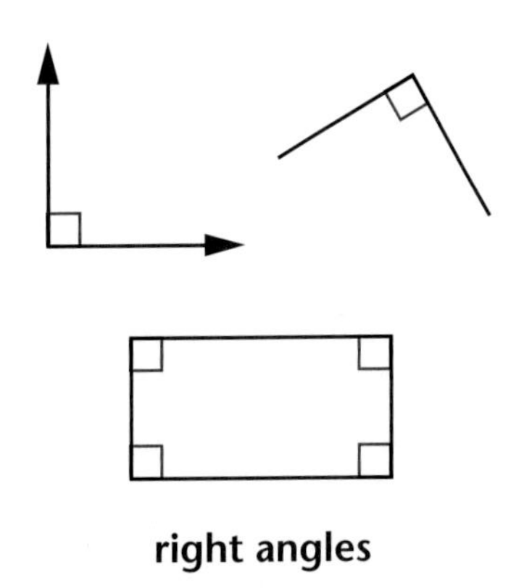
right angles

Right triangle A triangle that has one 90° angle.

right triangle

Round To adjust a number to make it easier to work with. Often, numbers are rounded to the nearest 10, 100, 1,000, and so on. For example, 864 rounded to the nearest hundred is 900.

Scale drawing A drawing that represents an actual object or region, but is a different size. Maps are scale drawings. Architects and builders use scale drawings.

Side (1) One of the rays or segments that make up an angle. (2) One of the line segments of a polygon. (3) One of the faces of a solid figure.

Solids Three-dimensional shapes, such as prisms, pyramids, cylinders, cones, and spheres.

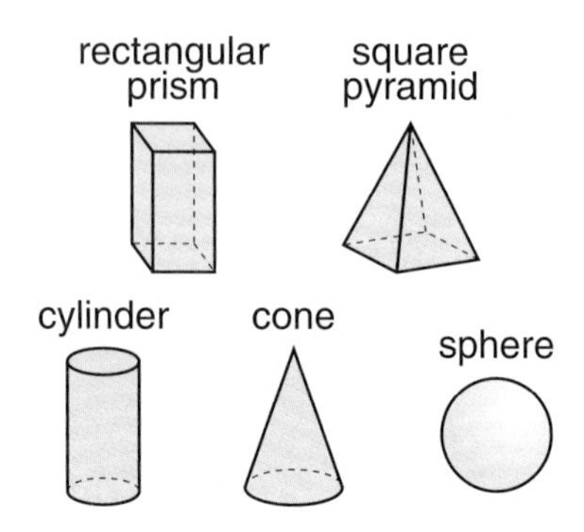

Sphere A solid with a curved surface that looks like a ball or globe. The points on a sphere are all the same distance from a point called the *center*. See *radius*.

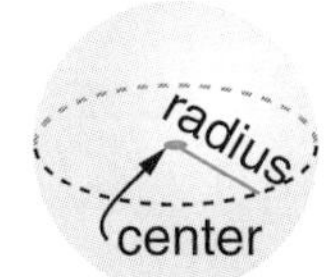

Square A *rectangle* whose sides are all the same length.

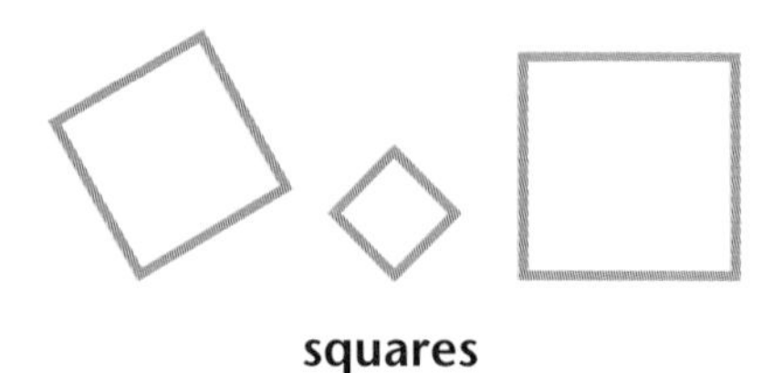

squares

Standard units Measurement units that are the same size no matter who uses them and when or where they are used.

Tally chart A chart that uses marks, called *tally marks,* to show how many times each value appears in a set of data.

Number of Pull-Ups	Number of Children
0	~~////~~ /
1	~~////~~
2	////
3	//
4	
5	///
6	/

Temperature A measure of how hot or cold something is.

3-dimensional (3-D) Having length, width, and thickness. Solid objects that take up space, such as balls, rocks, boxes, and books, are 3-dimensional.

Trade-first method One method for solving subtraction problems.

Trapezoid A 4-sided *polygon* that has exactly one pair of parallel sides.

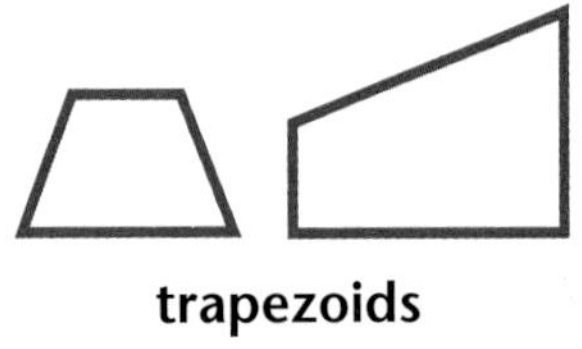

trapezoids

Triangle A *polygon* that has 3 sides and 3 angles.

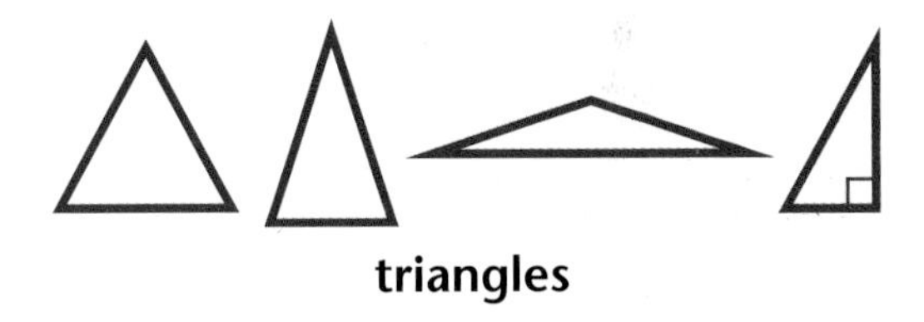

triangles

Turn-around facts Numbers can be added or multiplied in either order. $3 + 5 = 8$ and $5 + 3 = 8$ are turn-around addition facts. $4 \times 5 = 20$ and $5 \times 4 = 20$ are turn-around multiplication facts. There are no turn-around facts for subtraction and division if the numbers are different.

2-dimensional (2-D) Having length and width but not thickness. Flat shapes that take up area, but not space, are 2-dimensional. For example, rectangles, triangles, circles, and other shapes drawn on paper or a flat surface are 2-dimensional.

U.S. customary system A measurement system that is used most commonly in the United States. See the Tables of Measures on page 246.

Vertex (plural: **vertices**) A point where the sides of an angle, the sides of a polygon, or the edges of a polyhedron meet; any corner of a solid.

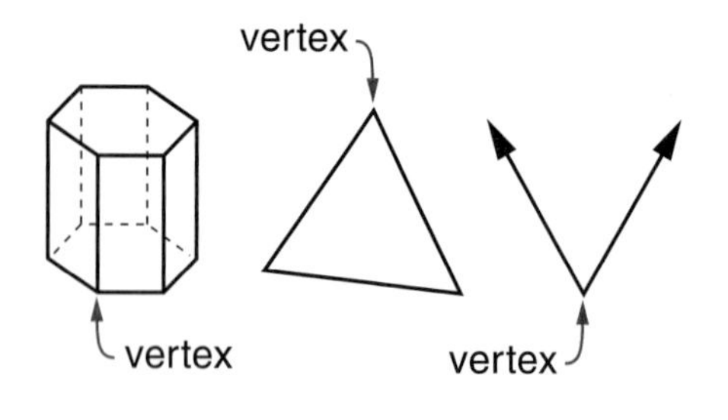

Volume The amount of space inside a 3-dimensional object. Volume is measured in cubic units, such as cubic centimeters or cubic inches. The volume or *capacity* of a container is a measure of how much the container will hold. Capacity is measured in units such as gallons or liters.

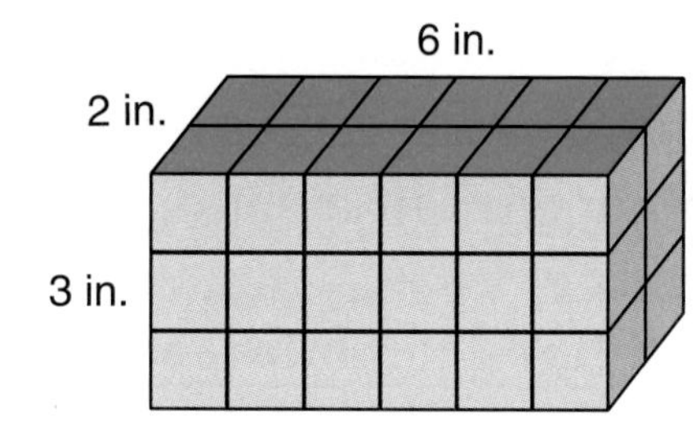

volume = 36 cubic inches

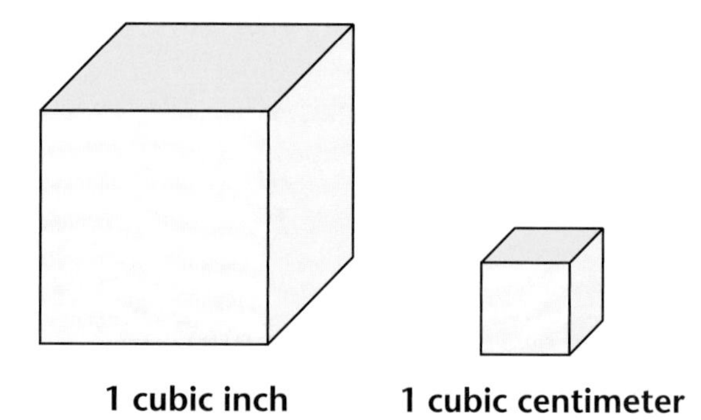

If the cubic centimeter were hollow, it would hold exactly 1 milliliter ($\frac{1}{1000}$ liter).

Weight A measure of how heavy something is.

Page 5

1. measure
2. code
3. count
4. location
5. comparison
6. measure
7. location
8. measure
9. count
10. comparison
11. location
12. measure
13. location
14. count
15. location
16. location

Page 9

1. a. 30
   b. 25
   c. 35
2. a.

		58	59
66	67	68	

b.

213		215	216
	224	225	

c.

	31		33	
40		42		44

Page 12

1. Number line: 30, $30\frac{1}{3}$, $30\frac{2}{3}$, 31, $31\frac{1}{3}$, $31\frac{2}{3}$, 32
2. Number line: 63, 63.2, 63.4, 63.6, 63.8, 64

Page 14

Sample answers: 3×4; 1.2×10; twelve; $3 + 3 + 3 + 3$; $12 \div 1$

Page 15

1. 9
2. **6** Sample answers:
   3×2 six
   $6 \div 1$
   $9 - 3$ $1 + 5$

Page 17

1. $(20 - 12) + 5 = 13$
2. $30 = 5 + (5 \times 5)$
3. $4 \times (7 + 14) = 84$
4. $16 = 2 \times (3 + 1) \times 2$

Page 21

1. 152, 162, 172, 182, 192, 202, 212, 222, 232, 242

2. 82,076

3. a. 4,789

b. 8,890

4. a. 6,671

b. 1,874

5. 753

Page 23

1. a. $\frac{5}{16}$

b. $\frac{2}{3}$

c. $\frac{4}{3}$

2. a. $\frac{1}{2}$; one-half

b. $\frac{4}{9}$; four-ninths

c. $\frac{6}{10}$; six-tenths

d. $\frac{8}{8}$; eight-eighths

Page 25

1. 1 inch

2. $1\frac{1}{2}$ inches

3. $1\frac{1}{4}$ inches

Page 29

Sample answers:
$\frac{2}{3}$; $\frac{4}{6}$; $\frac{8}{12}$; $\frac{40}{60}$; $\frac{200}{300}$

Page 32

1. $>$

2. $<$

3. $=$

4. $<$

5. $=$

6. $<$

7. close to 0

8. close to 1

9. close to 1

10. close to 0

11. close to 0

12. close to 1

Page 34

1. $\frac{80}{100}$; 0.80

2. $\frac{65}{100}$; 0.65

Page 36

1. 4.6

2. 0.6

3. 1.4

Page 51

1. $8 + 9 = 17$; $9 + 8 = 17$; $17 - 8 = 9$; $17 - 9 = 8$

2. $5 + 7 = 12$; $7 + 5 = 12$; $12 - 5 = 7$; $12 - 7 = 5$

3. $3 + 8 = 11$; $8 + 3 = 11$; $11 - 3 = 8$; $11 - 8 = 3$

4. $9 + 9 = 18$; $18 - 9 = 9$

Page 53

1. $8 \times 9 = 72$; $9 \times 8 = 72$; $72 \div 8 = 9$; $72 \div 9 = 8$

2. $5 \times 7 = 35$; $7 \times 5 = 35$; $35 \div 5 = 7$; $35 \div 7 = 5$

3. $3 \times 8 = 24$; $8 \times 3 = 24$; $24 \div 3 = 8$; $24 \div 8 = 3$

4. $9 \times 9 = 81$; $81 \div 9 = 9$

Page 55

1. a. $4 + 9 = 13$
   $9 + 4 = 13$
   $13 - 4 = 9$
   $13 - 9 = 4$

   b. $6 \times 8 = 48$
   $8 \times 6 = 48$
   $48 \div 6 = 8$
   $48 \div 8 = 6$

2. a.

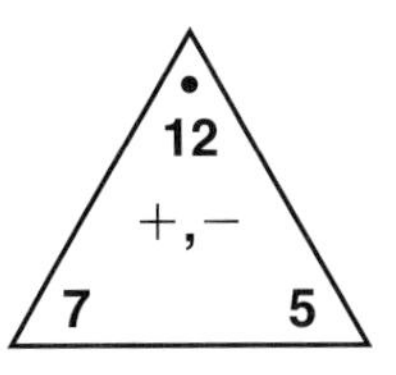

   b.

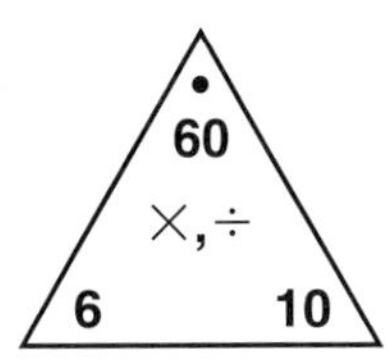

Page 57

1. 96
2. 122
3. 579
4. 801

Page 60

1. 38
2. 17
3. 176
4. 151

Page 63

1. 57
2. 172
3. 384
4. 372

Page 67

1. 32 chairs 

2. a. $4 \times 5 = 20$

   b. $3 \times 9 = 27$

Page 69

1. 222
2. 406
3. 11,018
4. 2,346

Page 72

1.

$4 \times 36 =$ 144

2. 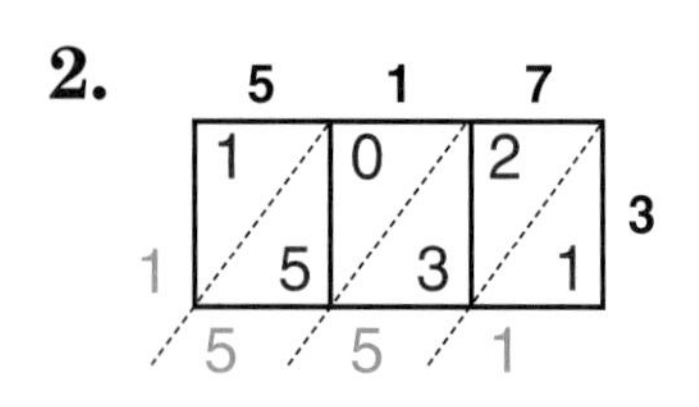

$3 \times 517 =$ 1,551

3. 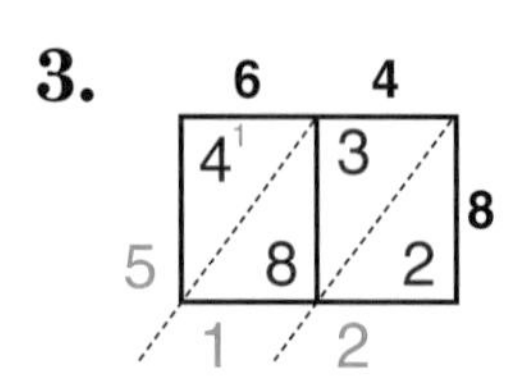

$64 \times 8 =$ 512

4. 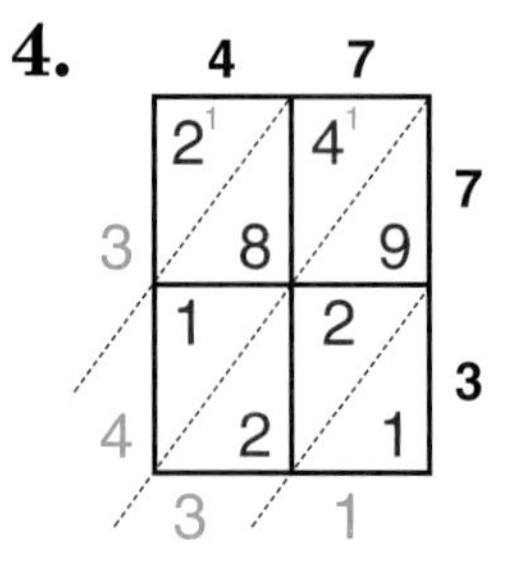

$47 \times 73 =$ 3,431

Page 76

1. 3
2. tomato juice; water

Page 78

1. a. Friday

   b. Tuesday

   c. 4 children

   d. 25 children

2. 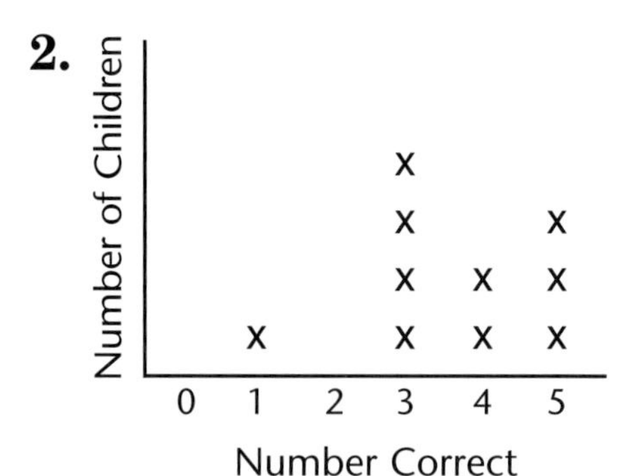

Page 82

1. Minimum: 23

   Maximum: 45

   Range: 22

2. a. Minimum: 0

Maximum: 10

Range: 10

b. 6

c. 7

3. 6

Page 85

1. 16

2. 8

Page 91

1. a. 60°F

b. 40°F

c. Saturday and Sunday

d. about 50°F because that is the median temperature (The mean temperature is about 50.3°F.)

e. 7

2.

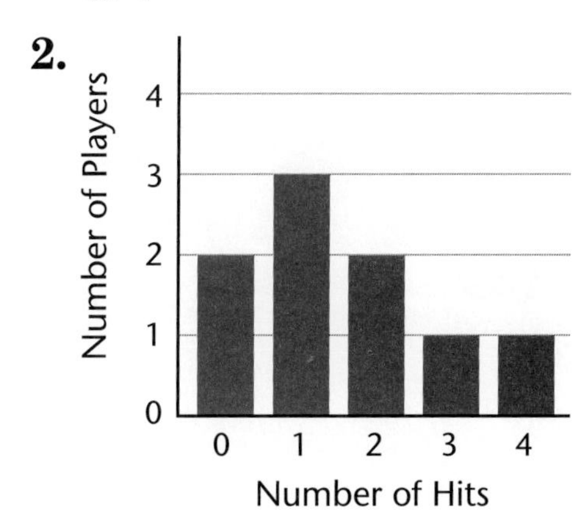

Page 103

1. a. hexagon

b. quadrangle or quadrilateral

c. decagon

d. octagon

e. dodecagon

2. Sample answers:

Page 111

1. 1 in.

2. $1\frac{1}{4}$ in.

3. $\frac{1}{2}$ in.

4. 1 cm

5. 3 cm

6. 2 cm

Page 121

triangle C

Page 123

1.

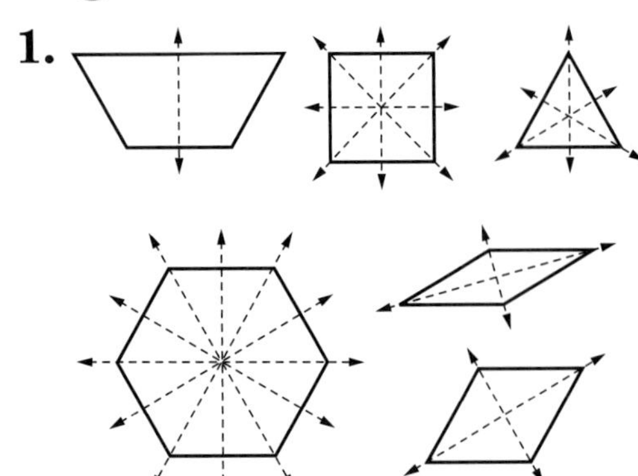

2. infinite; any line drawn through its center is a line of symmetry

Page 136

1. **a.** 100
   b. 10
   c. 1,000
2. millimeter; gram; meter; centimeter
3. **a.** 4-cm line segment

   b. 40-mm line segment

   c. Both line segments are the same length.

Page 142

1. 10,000 meters
2. **a.** 300 cm
   b. 350 cm
3. 4 cm
4. meters
5. **a.** 2 km
   b. 2,000 m
6. about 2.5 cm (25 mm), because a dime is about 1 mm thick

Page 145

1.

2. $2\frac{3}{4}$ in.
3. **a.** $\frac{1}{2}$ in.
   b. 1 in.
   c. $1\frac{3}{4}$ in.
   d. $2\frac{3}{8}$ in.
   e. $3\frac{1}{8}$ in.
   f. $3\frac{1}{2}$ in.

Page 149

1. **a.** 21
   b. 24
   c. 2
   d. 72

2. 63 in.

3. 12

4. 100 yd; 300 ft

5. 6 miles

Page 151

1. 15 ft

2. 60 mm

3. 39 in.

Page 153

1. about 21 mm

2. about 63 mm

3. dime

4. about 36 in.

Page 156

1. 14 sq cm

2. 27 sq in.

3. 1 square meter, because 1 meter is longer than 1 yard (see page 142)

Page 159

1. 1 cubic meter, because 1 meter is longer than 1 yard (see page 142)

2. a. 9 cu cm

b. 10 cu in.

c. 60 cu ft

Page 161

1. a. 20

b. 4

c. 3

2. about 40 cups

Page 163

Yes. 1 pound = 16 ounces.
1 ounce equals about 30 grams.
So 1 pound equals about
16×30 grams, or 480 grams.

Page 164

1. 1 gram; 1 ounce; 1 pound; 1 kilogram

2. a. 1

b. 1

c. 1,000

d. 0.6

e. 160

f. 4

3. the precision of the scale

Page 173

1. **a.** 100
   b. 32
   c. 98.6
   d. about 20
   e. −18
2. 38°F
3. 80°C
4. 38°F

Page 175

1. before noon and after noon
2. 12:00 A.M.; 2:55 A.M.; 4:15 A.M.; 10:50 A.M.; 12:00 P.M.; 3:05 P.M.; 7:30 P.M.; 9:45 P.M.
3. 10 years
4. 300 years
5. 24
6. 28
7. 3,600 seconds

Page 177

1. January; March; May; July; August; October; December
2. Thursday
3. May 26
4. Friday, June 6

Page 179

1. March 21 to June 20 north of the equator; September 22 to December 21 south of the equator
2. Houston has about 4 more hours of sunlight than Seward on December 22.

Page 181

1. **a.** *B*
   b. *J*
2. **a.** (2,2)
   b. (5,7)
3. **a.** (1,2)
   b. (6,4)
   c. (5,0)
   d. (3,1)
   e. $(4, 5\frac{1}{2})$
   f. $(1\frac{1}{2}, 4\frac{1}{2})$

Page 194

1. 80
2. 30
3. 90
4. 600
5. 4,000
6. 300 + 50 = 350

Page 197

1.

2.

Page 229

1. croquet ball, basketball, bowling ball
2. bowling ball
3. yes, except for croquet balls

Page 238

1. about 500 (2,562 – 2,072)
2. about 600 (896 – 272)

Page 239

1. A, E, N, O, and T
2. J, K, Q, X, and Z
3. 82 + 130 + 65 + 80 + 25 = 382

Page 242

1. Theodore Roosevelt
2. Ronald Reagan

Page 252

1. Each friend got 5 pennies.
2. The perimeter of the rectangle is 50 cm.
3. Each marker cost 14¢.
4. The numbers are 3 and 4.
5. There are 14 boys in the class.

Page 257

Total	
80	
Part	Part
36	?

Ulla read for 44 minutes.

Page 266

1. 7
2. 12
3. 30
4. 8
5. 5, 8, 11, 14, 17, 20
6. 10, 8, 6, 4, 2, 0

D

SRB
347

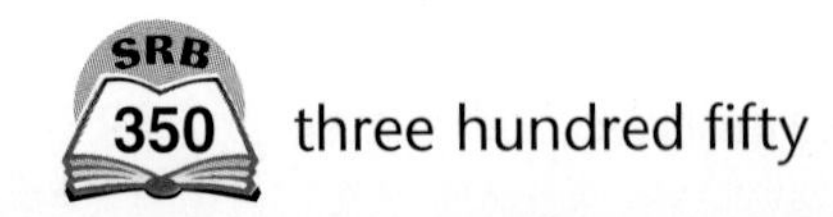

T

Photo Credits

©Theo Allofs/Corbis, p.127 *top*; ©Bettmann/Corbis, p.185 *right*; p.186 *top left*; ©Jonathan Blair/Corbis, p.209 *top*; ©Ed Bock/Corbis, p.235; ©Tom Brakefiel/Getty Images, p.207 *bottom*; ©The British Museum, p.46 *top*; ©José Manuel Sanchis Calvete/Corbis, p.187 *top*; ©Laurie Campbell/Getty Images, p.206 *bottom*; ©Steve Carp, pp.133, 150; ©Ralph A. Clevenger/Corbis, p.174 *bottom*; ©Clouds Hill Imaging Ltd./Corbis, p.125 *center*; p.129 *top right*; ©Corbis, p.45 *center left*; p.126 *top* and *center*; p.130 *top left*; p.183 *bottom*; ©Richard Cummins/Corbis, p.129 *center*; ©DK Limited/Corbis, p.209 *center*; ©Bob Daemmrich/Photo Edit, p.40; ©W. Treat Davidson/Photo Researchers, Inc., p.219 *bottom*; ©Tim Davis/Getty Images, p.128 *center inset*; ©Araldo de Luca/Corbis, p.45 *top*, p.186 *bottom right*; ©Goupy Didier/Coribs, p.207 *top*; ©Don Farrall/Getty Images, p.233; ©Felix/Zefa/Corbis, p.130 *top right*; ©Peter M. Fisher/Corbis, p.188 *top*; ©Tim Flach/Getty Images, Cover; ©D. Fleethan/OSF, p.218 *top left*; ©Natalie Fobes/Corbis, p.206, *bottom inset*, p.208 *bottom inset*; ©Tony Freeman/Photo Edit, p.78; ©GC Minerals/Alamy, p.129 *top left*; ©Jeff Greenberg/Index Stock, p.104 *bottom*; ©Christel Gerstenberg/Corbis, p.46 *bottom left*; ©Getty Images, Cover, *bottom left*; v, vi, vii; p.127 *center*; p.128 *bottom*; ©Todd Gipstein/Corbis, p.47 *center*; ©Darrell Gulin/Corbis, p.130 *bottom*; ©Handout/Reuters/Corbis, p.45 *center right*; ©Martin Harvey/Corbis, p.206 *top inset*; ©Inglolf Hatz/Zefa/Corbis, p.186 *top right*; ©Jason Hawkes/Corbis, p.104 *bottom right*, 184 *top*; ©Sharon Hoogstraten, pp.25, 142, 144, 145, 152, 160, 172, 216 *bottom right*, 217 *top right*, 251, 255, 307; ©Kit Houghton/Corbis, p.207 *bottom inset*; ©Mimmo Jodice/Corbis, p.104 top; ©Dmitry Kessel/Stringer/Getty Images, p.185 *top*; ©Nick Koudis/Getty Images, p.45 *bottom*; ©LWA-JDC/Corbis, p.43 *bottom*; ©Frans Lanting/Corbis, p.130 *bottom inset*; ©David Lees/Corbis, p.44 *top*; ©Charles and Josette Lenars/Corbis, p.113; ©George D. Lepp/Photo Researchers, p.218 *second from top*; ©Library of Congress, p.242; ©Renee Lynn/Getty Images, p.128 *center*; ©Tom McHugh/Photo Researchers, Inc., p.209 *bottom*; ©Jean Mahaux/Getty Images, p.192; ©Leo Meier/Australian Picture Library/Corbis, p.206 *top inset*; ©Lawrence Migdale/Getty Images, p.82; ©Roy Morsch/Corbis, p.208 *bottom*; ©NASA/Corbis, pp.118 *bottom center*, 128 *top*; ©C. Netscher/Corbis, p.186 *bottom left*; ©Eric Nguyen/Jim Reed Photography, p.129 *bottom*; ©R. Andrew Odum/Peter Arnold, Inc., p.218 *bottom left*; ©Darren Padgham/Getty Images, p.113; ©photolibrary.com, p.48 *right*; ©Judd Pilossof/Getty Images, p.76 *center*; ©Carl and Ann Purcell/Corbis, p.47 *top*; ©H. Reinhard/Zefa/Corbis, p.126 *bottom left*; ©Roger Ressmeyer/Corbis, p.125 *bottom*; p.183 *right*; ©Joel W. Rogers/Corbis, p.188 *bottom*; ©Norbert Rosing/Getty Images, p.210 *right*; ©Anderson Ross/Getty Images, p.3; ©Kjell B. Sandved/Visuals Unlimited, p.219 *third from top*; ©E. Sauer/Zefa/Corbis, p.130 *center*; ©Scala/Art Resource, NY, p.46 *bottom right*; ©Kevin Schafer/Corbis, p.127 *bottom*; ©Science Museum, p.48 *left*, p.185 *bottom*; p.187 *center* and *bottom*; ©Ariel Skelley/Corbis, p.43 *center left*; ©Don Smetzer/Photo Edit, p.40; ©Paul A. Souders/Corbis, p.205 *top*; p.208 *top*; ©Ruben Sprich/Reuters/Corbis, p.175; ©SSPL/The Image Works, p.72; ©Chase Swift/Corbis, p.183 *top*; ©Steve Terrill/Corbis, p.128 *bottom inset*; ©Tom Thistlethwaite/Corbis, p.113; ©David Turnley/Corbis, p.47 *bottom*; ©Vanni Archive/Corbis, p.184, *center*; ©Kennan Ward/Corbis, p.210 *left*; ©Michele Westmoreland/Corbis, p.44 *bottom*; ©Staffan Widstrand/Corbis, p.174 *top*.